REF

Tillman, David A.

Incineration of
municipal and
hazardous solid
wastes

$42.95

DATE			

DISCARD

Incineration of
Municipal and Hazardous
Solid Wastes

Incineration of Municipal and Hazardous Solid Wastes

David A. Tillman
Envirosphere Company
Bellevue, Washington

Amadeo J. Rossi
Envirosphere Company
Bellevue, Washington

Katherine M. Vick
Envirosphere Company
Bellevue, Washington

Academic Press, Inc.
Harcourt Brace Jovanovich, Publishers
San Diego   New York   Berkeley
Boston   London   Sydney   Tokyo   Toronto

Academic Press Rapid Manuscript Reproduction

Academic Press, Inc.
San Diego, California 92101

United Kingdom Edition published by
Academic Press Limited
24–28 Oval Road, London NW1 7DX

Library of Congress Cataloging in Publication Data

Tillman, David A.
    Incineration of municipal and hazardous solid wastes / David A.
Tillman, Amadeo J. Rossi, Katherine M. Vick.
        p.   cm.
    Includes bibliographies and index.
    ISBN   0-12-691245-9   (alk. paper)
    1. Incineration--United States. 2. Refuse and refuse disposal-
-United States. 3. Hazardous wastes--United States.   I. Rossi,
Amadeo J. II. Vick, Katherine M. III. Title.
TD1062.T55   1989
628.4'457--dc19   88-39132
                                                         CIP

Printed in the United States of America
89 90 91 92   9 8 7 6 5 4 3 2 1

For Alexis,
Anthony Joseph,
Coogan, and Darek

Contents

Chapter V
Fundamentals of Solid Hazardous Waste Combustion

Chapter VI
Permanent Solid Hazardous Waste Incineration Systems

Chapter VII
Mobile, Transportable, and Developing Incineration Systems

Chapter VIII
Controlling Products of Combustion

Preface

Solid waste management has become one of the central urban issues confronting the United States and the Western world. Solid wastes are generated daily by households, hospitals, commercial establishments, industries, governmental operations, and virtually every element of society. Much of this solid waste is simply trash: rubbish or garbage. A significant quantity of this solid waste is hazardous. All of this solid waste must be managed—and ultimately disposed of.

The hundreds of thousands of tons of solid waste generated annually are the result of ordinary manufacturing, distribution, purchasing, use, and disposal activities of industrialized societies. These solid wastes result from an economy designed to deliver the maximum quantity of product to the industrial, commercial, and household consumer at the minimum cost. They include paper and plastics, wood, rubber and leather, old clothing, caustic compounds for cleaning ovens, reductants for polishing silver and cleaning copper, and a host of other products.

An array of waste management and disposal strategies has emerged over the past hundred years, including landfilling, recycling, composting, and incineration. Landfilling, or land disposal, has been a traditional method for waste management since time began. Landfills today are sophisticated, technological repositories for raw waste and the results of waste processing (e.g., incinerator ash and slag). They may be designed for safe waste management and disposal over many years. The availability of substantial, suitable acreages of land is necessary in order for the land disposal system to be an exclusive, cost effective, waste management strategy. Recycling has long been a waste reduction strategy. Entire industries have their origins in recycling, including the pulp and paper industry and the electric furnance "mini-mills," which produce increasing quantities of steel in the United States. Recycling reached perhaps its highest levels during World War II, when the combination of raw material scarcities and patriotism created both the demand and the supply components necessary for this waste management strategy. Recycling industries, dealing largely with industrial waste, regularly contribute more than $4 billion annually to the U.S. Gross National Product. Recycling industries have achieved significant technological breakthroughs in the past twenty years, including the development of the automobile shredder and the recycling of lead batteries. Composting is a tradition in many agricultural areas and also has tangential urban traditions such as feeding garbage to pigs on the county farm. Incineration has an equally long tradition as a volume reduction technology, although its roots are sometimes far from illustrious.

Today, landfilling, recycling, composting, and incineration remain the critical elements of municipal and hazardous solid waste management strategy. Further, all of these elements have received substantial infusions of new technology. The modern municipal waste incinerator, for example, is a sophisticated system designed for complete, high temperature destruction of waste. It includes significant postcombustion controls for the removal of undesirable airborne emissions. The modern municipal waste incineration plant may include some materials recovery as well as energy recovery. It also includes ash control and management systems. Similarly, the hazardous waste incinerator is a sophisticated system designed for complete destruction of the toxic components in the waste fed to it. Like the municipal waste incinerator, it also includes sophisticated air quality control systems and ash management systems.

Despite the many technological advances in solid waste combustion, an increasing resistance among many public officials and some vocal segments of the population inhibits the use of incineration as one rational waste management strategy. This resistance is a result of many forces, including memories and experiences with earlier incineration technologies that are now obsolete, increasing knowledge concerning potential problems that could be caused by excessive emissions from incinerators, and a general public skepticism concerning technological solutions to problems (e.g., nuclear power). This resistance has become grist for many reporters in the news media. Further, this resistance impedes implementing the recent gains in knowledge and the recent technological developments associated with incineration systems. Combustion of various solid wastes is entering an exciting era of technical improvements designed to further this particular weapon in the waste management arsenal.

This book is designed to address many of the developments in applying the combustion process to the incineration of solid municipal and hazardous wastes. The book establishes the waste management context. It examines the fundamental scientific basis for combustion of municipal and hazardous wastes. It considers the processes now available for such incineration. It concludes by discussing the air quality control systems available.

The authors performed considerable work evaluating municipal and hazardous waste management issues, combustion principles, and combustion systems. This work was performed for many clients of Envirosphere Company throughout the United States. It has required site visits to many municipal and hazardous waste incineration systems throughout the United States and Europe. At the same time, it has involved the development of combustion models for computerized analysis of problems. The research for this book took several years. Further, the book is somewhat of an extension of our work, which began in the middle and late 1970s, evaluating combustion and energy recovery aspects of municipal waste and biomass. In reality, this work is not complete, as we continue to progress in these essential waste management arenas.

In this book we summarize combustion theory as it applies to full scale systems. The book is an attempt to bring theory and commercial practice together. This effort resulted in numerous anomalies, including variations in units of measure. At times metric units appeared more appropriate. For the most part, however, the information was developed in English units to meet the needs of projects, clients, and other colleagues. Other anomalies include some of the descriptive approaches taken.

The writing of this book has required the assistance of numerous individuals, whose invaluable contributions are acknowledged below. Junior colleagues, including Tim Gould, Liz Balzer, and Tracy Wegehaupt, provided assistance with figures, references, and many details. Professional associates provided invaluable insights and ideas, contributing to the overall project. Further, several professional colleagues contributed significantly to document review. These individuals include Terrill Chang, Vice President of SCS Engineers; Harry Hall, Manager of Resource Recovery for King County, Washington; Jim Kerstetter of the Washington State Energy Office; Gerald Smedes of Rabanco Companies; Bob Ferraro of the John Zink Company; Jim Nicotre and Vic White of Environmental Elements Company; Steve Anderson of Roy F. Weston Corporation; and Mo Massoudi and John Osterle of Ebasco Services Incorporated. Numerous individuals provided plant tours at such facilities as McKay Bay, Tampa, Florida; Pinnellas County, Florida; Nashville, Tennessee; Sumner County, Tennessee; Baltimore, Maryland; North Andover, Massachusetts; Columbus, Ohio; Hartford, Connecticut; Madison, Wisconsin; Tacoma, Washington; SAKAB, Nortorrp Sweden; HIM, Beibesheim, West Germany; and Coffeyville, Kansas. Lars Olsson of the Swedish Trade Council facilitated plant tours and information gathering in numerous facilities in Sweden. The list of vendors providing supporting information is legion. Research support with test information, data, and ideas came from Hazen Research of Golden, Colorado; Dr. David Pershing, Dean of Engineering at the University of Utah; and Dr. W. Randall Seeker of EER Corporation. We also gained support from superiors at Envirosphere Company. Millie Tillman assumed responsibility for typing the final manuscript.

Incineration of Municipal and Hazardous Solid Wastes, then, is an examination of the principles and processes associated with the combustion of many materials discarded by society. It is intended to provide some insight into the process of combustion and the application of that process to one of the leading problems confronting our everyday lives. If focuses on first principles and the application of those principles by technology.

David A. Tillman
Amadeo J. Rossi
Katherine M. Vick

Chapter I

WASTE GENERATION IN THE UNITED STATES

I. INTRODUCTION

Incineration, the thermal destruction of organics by
combustion, or high temperature oxidation, is one of the many
techniques used for the treatment of municipal solid wastes
(MSW) and hazardous wastes. As such it is one potential element in any overall waste management system. The basic components of any waste management system include: (1) generation;
(2) collection and concentration; (3) treatment; and (4) final
disposal. Techniques for waste management associated with generation focus on waste minimization and recycling. Collection
and concentration involves packer trucks and transfer stations
for MSW, and collection systems and blending facilities for
hazardous waste. Final disposal typically involves landfilling
and related techniques. Incineration is considered a treatment
technology for both MSW and hazardous waste. Additional treatment technologies for MSW include composting while other technologies to treat hazardous waste include permanganate oxidation; neutralization; and stabilization by cement, pozzolon,
or polymeric means.
Incineration has several distinctive characteristics. For
the most part, it dramatically reduces the volume of waste to
be landfilled. Further, incineration chemically transforms
MSW and hazardous waste and, when combined with stabilization,
can produce a material which is relatively benign in the landfill. Incineration is also among the most capital-intensive
solutions to both MSW and hazardous waste disposal. Consequently, its economics are driven as much by capital recovery
as by labor and maintenance costs.
Historically, the incineration process has been applied to
a wide variety of wastes, and with varying degrees of success.
Older, inadequate incinerators have appeared to cause more
problems than they have solved. High technology incineration-
related approaches to thermal destruction of MSW and hazardous
waste have been introduced from time to time, and many have
become well publicized and costly failures (i.e., municipal
waste pyrolysis and liquefaction). Consequently, incineration
of MSW and hazardous waste has become controversial with both
the regulatory agencies and the general public.

Technically, however, incineration is simply the process of thermally oxidizing various wastes. It offers distinct applications to MSW and hazardous waste but also has certain limitations. Incineration is not *the* answer; rather, it is one waste management technique which can be used either to compete with or, more properly, to complement such strategies as waste minimization, recycling, stabilization, and well managed landfilling. In order to consider the incineration process, the following issues must be addressed: (1) the nature of the problems in managing MSW and hazardous waste in the United States and in industrialized economies; (2) the fundamentals of the combustion process applied to MSW and hazardous waste; and (3) the technologies available for effective incineration of MSW and hazardous waste, including the technologies available for control of airborne emissions resulting from combustion. These issues form the basis for this book.

A. Overview

Over 10,000 landfills of MSW dot the landscape of the United States (see Fig. 1). Many landfills are no longer used,

FIGURE 1. Large, well managed, sanitary landfill in Southern California.

having been closed for environmental and other reasons. Over
one-third of the solid waste landfills still operating will
reach design capacity and the consequent ability to accept
waste within five to seven years [9]. Even now, some communities have no place to dispose of their MSW and are facing
crises of monumental proportions. For example, in 1978 there
were over 400 landfills in New Jersey. Today, less than 100
are operating, and of those, two receive most of the garbage
[9]. Since 1983 New York has closed more than 200 landfills,
and the state expects to exhaust the remaining landfill
capacity in ten years [9]. Even with the renewed interest in
recycling, many communities are facing difficult decisions
about dealing with their garbage, because recycling is not a
total solution.

The U.S. leads the world in per capita production of MSW.
Each U.S. citizen generates four to five pounds (1.8-2.3 kg)
of garbage each day. This society as a whole generates over
150 million tons (136 million tonne) [12] of garbage each
year. By the year 2000, the U. S. Environmental Protection
Agency (EPA) has predicted that this figure will increase about
20% [12].

In addition to the problems with MSW landfills, tens of
thousands of sites throughout the country are contaminated
with hazardous or toxic wastes. The General Accounting Office
estimates that there may be as many as 425,000 waste sites
that are potential hazards [6]. EPA has currently listed only
27,000 of these as possible threats, and has investigated
approximately 23,000 sites [6]. By far, the majority of sites
are found in the industrial areas of the country.

Only recently have the pressures of rapidly diminishing
landfill space, high disposal costs, and environmental problems
caused a serious exploration of alternatives in solid waste
management and disposal. Modern societies have demanded numerous convenience and throw-away goods to sustain the fast
pace of our industrial society lifestyle. Industry has
responded with products ranging from styrofoam cups to panty
hose. The U.S. is not alone in contributing to and dealing
with the problem. Many European countries, as well as Japan,
began facing this problem decades ago, and many selected
incineration as part of the solution. As we all proceed to
clean up MSW and hazardous waste sites, and to improve the
management of wastes, the issues of cost as well as appropriate technology loom on the horizon.

This chapter initiates the analysis of the following
questions. In what application is the combustion of MSW and
hazardous waste a practical management tool? Where are the
principles of combustion and how do they apply to MSW and
hazardous waste? What are the combustion technologies, and

can they meet strict environmental standards for emissions?
To establish a basis from which to address those questions in
the remainder of the book, this chapter first discusses the
current status of waste generation. Next, MSW and hazardous
waste are reviewed. The chapter then covers the generation
of each waste from an historical perspective leading to the
focus on the current status.

B. Current Patterns of Waste Generation

     Historically, wastes have always created a disposal
problem. For centuries, most wastes were decomposable
organics. As industrialized societies have grown, significant
changes in wastes and in the methods for dealing with those
wastes (Table I) have become apparent and necessary. Most
wastes no longer are biodegradable. The transition from an
agricultural society to an industrial and now to a post-
industrial society has increased the quantity of complex and
nonbiodegradable wastes which are far more difficult and chal-
lenging to manage.
     During the past 40 years the production, distribution, and
use of potentially hazardous substances have increased dramatically. Growing populations have demanded more products and
services which have resulted in the accelerated manufacture of
synthetic materials. For example, the percentage of plastics
has increased significantly in the MSW stream over the past

TABLE I. Municipal Waste Stream Components

	Percent of municipal waste stream		
Materials	1960	1984	2000
Paper	32.1	37.1	41.0
Yard wastes	20.3	17.9	15.3
Glass	8.4	9.7	7.6
Metals	13.7	9.6	9.0
Food	14.6	81.0	6.8
Plastics	0.5	7.2	9.8
Other	10.4	10.4	10.5
Total	100.0	100.0	100.0

Sources: [3, 11].

24 years. With an increase from zero to about 9% of the MSW
stream, plastic clearly reflects the changing consumer habits
and life styles of society [3]. In addition, societies are
disposing of increasing quantities of solid waste, due in part
to population growth, to increased affluence, and technological changes. Coupled with the consumption of significant supplies of raw materials, and an economy that demands growth,
the long-term cost is ultimately damage to the environment.

An example of an MSW stream component which is both a
benefit and a problem is the nonbiodegradable six-loop plastic
holder for beverages. The loops provide a convenient, low-
weight method for consumers to carry goods. However, they
have been banned in many states because the loops have caused
numerous marine mammals and birds to die from ingestion or
entanglement. Another example from the hazardous waste perspective is groundwater contamination resulting from the use
of trichloroethylene (TCE) and other commonly used cleaning
solvents. From the early 1940s through the mid-1960s, many
military bases used trichloroethylene as a degreaser. It was
a "miracle" cleaning agent because it did its job so well.
TCE and analogous compounds have been used extensively by the
dry cleaning industry. However, the chemical has seeped into
groundwater and contaminated drinking water supplies throughout the United States and other industrialized countries.

How best to handle the changing waste stream safely and
economically is a key question facing decisionmakers. New
products have significantly altered the composition of wastes
and have demanded new techniques to properly control hazards
associated with their disposal. The role of the public in the
decisionmaking process has also become a key concern in the
handling of wastes. Given the increasing public interest in
activities which affect the environment, and the regulatory
support for public involvement in environmental decisionmaking, the public has taken on the role of a "shadow regulator."
No decisionmaker sensitive to the concerns and desires of the
public can ignore the increasing strength and commitment of
the public in the unofficial but strong role of regulator and
proponent of a clean environment. As many waste experts have
learned, the focus of public concern may emerge as the "not in
my backyard" (NIMBY) syndrome when dealing with MSW and
hazardous waste. Further, the memories, perceptions, and
intuitive reactions of the public may be directly opposite to
the findings and knowledge of the scientific and engineering
communities.

II. MUNICIPAL SOLID WASTE (MSW) MANAGEMENT

Historically, most MSW has been disposed of through land-filling. This disposal method proved to be efficient and inexpensive until changing waste compositions and past disposal methods resulted in groundwater contamination and other problems. These problems have resulted in landfill costs increasing over 300% in some areas [4], with predictions even higher in the future because of potential remediation and the need to provide long-term monitoring. With the rise in the cost of disposal, however, many municipalities are seeking other methods and solutions to managing their wastes. The following section considers the amounts and composition of MSW generated, its disposition, applicable regulations, and issues associated with disposal.

A. MSW Generation and Composition

The United States generates approximately 450,000 tons per day of MSW [15]. This amounts to an average of 1 ton per person per year. Although composition varies from city to city and also according to seasons, more than two-thirds of the MSW (on a weight basis) is comprised of organic or combustible materials (i.e., paper, wood, food, yard clippings) [15]. The remaining one-third is primarily metals, glass, and dirt [15]. Approximately 90% of the MSW is disposed of by land burial [9]. European societies generate less paper and substantially more quantities of plastics than the United States (see Table II). Much of the European waste is no longer directed to landfills but rather is combusted in incinerators to reduce volume prior to final disposal.

While paper is the largest portion of the MSW stream and is predicted to continue growing, plastic comprises the fastest growing segment. For example, in California, the amount of plastic in the waste stream has risen from 3% in 1980 to nearly 7% in 1986. Implications resulting from this compositional change include a higher portion of the waste that will be easily combustible since both paper and plastics have high heat values.

B. Issues and Trends in MSW

Although landfilling is still employed as both a treatment and a disposal technique, trends in managing MSW are shifting from landfilling as a primary treatment technique to favor a broader slate of management methods including recycling and incineration along with land disposal of the

TABLE II. Approximate Composition and Generation Rates of Solid Waste in Western Europe (Values in Weight Percent)

Component	UK	France	Nether-lands	West Germany	Switzer-land	Italy	U.S.
Garbage (organics)	27	22	21	15	20	25	12
Paper	38	34	25	28	45	20	50
Fines	11	20	20	28	20	25	7
Metal	9	8	3	7	5	3	9
Glass	9	8	10	9	5	7	9
Plastics	2.5	4	4	3	3	5	5
Miscellaneous	3.5	4	17	10	2	15	8
Average water content (%)	25	35	25	35	35	30	25
Caloric content Btu/lb	4,200	4,000	3,600	3,600	4,300	3,000	5,000
Generation lb/cap/yr	700	600	455	770	550	465	1,800

Source: Pavoni et al. [10].

residue.   In the overall management of wastes a variety of
techniques must be used to complement one another.   These
techniques include reducing the amount of MSW generated;
recycling appropriate components of the waste stream; incinerating and recovering energy from nonrecyclables; and landfilling the remaining MSW and ash.

Recent public sentiment and legislation at both the federal
and local levels have encouraged the use of reduction as a
tool to manage MSW; that is, reducing or minimizing the amount
of MSW generated.   This strategy addresses actions such as
legislation to restrict packaging or to encourage consumers
to purchase only those goods which can be reused--thereby
reducing the amount of waste generated.   Interest in recycling
has also been renewed as a viable component of the MSW management system.   Interestingly, the highest rates of recycling
were achieved during World War II due to stable material supply and, more importantly, a substantial demand for recovered
materials.   After World War II, recycling steadily decreased
due to social and economic factors.   Even when recycling
became a practice, industries were still creating substantial
economic benefits from it.

The recycling industry has always been a $4,000,000,000/yr
economic sector [8] supported in part by industries such as
iron and steel, stainless steel, and pulp and paper.   Pulp and
paper began as a recycling industry and converted to the use
of wood fiber for raw material between 1850 and 1920.   A wide
variety of industries traditionally have supported recycling,
ranging from detinning processes for cans to auto shredding.
Recycled materials result from both industrial scrap and post-
consumer wastes such as residential waste.   Recycling of
household refuse relies on the participation of individuals,
however.   The long-term success of a general program based
upon individual participation for nonpatriotic reasons remains
undemonstrated and is far from assured.

Communities have responded to the close of landfills and
the proposed construction of incinerators with pressure for
more recycling first.   Although some estimates indicate that
the U.S. is only recycling about 10% of all MSW [13], many
experts suggest that an aggressive recycling campaign could
reach substantially higher levels within a relatively short
period of time.

Incineration provides another waste management option and
many communities in the U.S. have selected various modern
incineration technologies for the disposal of their waste
(see Fig. 2).   The incineration technologies employed range
from commercial scale mass-burn facilities to a variety of
refuse-derived fuel processes.   Many of these incineration
facilities have state-of-the-art air quality control systems.

FIGURE 2. The Nashville Thermal Transfer Co. incineration plant, Nashville, Tennessee. It was among the first of the modern incinerators built in the United States.

The main concern about incineration expressed by the public, however, is airborne emissions and ash disposal. Past exper- iences of older incinerators have been marred by accidents and poor emission controls. Even though new control technologies and monitoring programs may provide for sufficient protection of public health, environmental concerns and pressures have resulted in the cancellation or reduction in size of many proposed incineration facilities.

Landfilling is still the final technique for waste dis- posal but neither the method nor the costs for landfilling have remained static. Although the USEPA has yet to propose its new landfill facilities rules, these regulations are likely to impose strict requirements for new and existing sites. Such requirements will result in a substantial increase to the cost of landfilling MSW. In addition, problems abound with old landfills and the public has become more aware of those problems, such as contaminated water supplies from leachate or diminished land values from methane gas.

C. Legislation Affecting MSW Management

In addition to the Federal Government, most state and local governments have legislated certain MSW amangement activities. The State of New Jersey recently passed a law requiring mandatory recycling in that state. The City of Berkeley, California, has mandated that no styrofoam food packaging would be purchased by the City and instructed fast- food outlets to begin complying with the ban on certain plas- tic products. As early as 1973, the states of Oregon and Vermont passed "bottle bills" to help manage the problem of litter along the highways. The legislation required deposits on all beverage containers. Since then at least a dozen other states, have passed legislation which requires deposits on beverage containers. Even local health departments have played key roles in the management of MSW because many are responsible for issuing permits certifying the safety and pro- tection of the public health related to a facility.

D. Future Trends in MSW Management

Trends in MSW point to ever-increasing amounts of waste being generated. Some predictions indicate an increase from the 150 million tons (136 million tonne) per year of MSW now generated in the U.S. to more than 250 million tons (227 mil- lion tonne) per year in 2000 [3]. As landfilling costs increase, and as recycling proves incapable of managing the entire MSW stream, incineration will likely be encouraged and

promoted. Although only 4% of the MSW presently generated in
the U.S. is being incinerated, predictions to the year 2000
range as high as 25% of MSW will be incinerated as a disposal
method [6].

III. HAZARDOUS WASTES MANAGEMENT

According to the Resource Conservation and Recovery Act
(RCRA), a waste is defined as hazardous if it exhibits properties of ignitability, corrosivity, reactivity, or toxicity.
Additionally, a waste or waste stream is considered hazardous
if it has been specifically listed in the federal regulations
or is a mixture of a listed hazardous waste and nonhazardous
waste. In general, Congress has defined hazardous wastes as
those discarded materials which may threaten human health or
the environment when improperly disposed.

A. Generation and Composition of Hazardous Wastes

In 1981 the USEPA estimated that 290 million tons (263
tonnes) of hazardous waste are generated in the U.S. anually
[6]. A recent study indicates that more than 300 million tons
of hazardous wastes were generated in 1987, and that by 1990
more than 250,000 sites will be generating toxic wastes.
Even the U.S. household contributes to the hazardous waste
problem by generating about 10 pounds of hazardous wastes
each year. These items include cleaning solvents, paints,
motor oil, garden chemicals, and other products.
Hazardous wastes may be in any of the following forms:
solids, liquids, sludges, or contained gases. These wastes
are generated by a variety of sources, including industry,
the military, hospitals, research institutions, schools,
businesses, and households. However, because most manufacturing processes have the potential to create some hazardous
waste, industry has been identified as the largest assembly
of generators. The predominant industrial sectors associated
with hazardous waste generation are identified in Table III.
Since the passage of the National Environmental Protection
Act (NEPA) in 1970, federal and state regulations have
resulted in significant progress to control, manage, and dispose of hazardous wastes. It is estimated that industry and
the federal government spend more than $70 billion per year
to manage and dispose of hazardous wastes [6]. According to
USEPA, only one out of every six generators manage their
hazardous wastes on site. Most ship their wastes off site to
be treated, stored, or disposed [6].

TABLE III. Number of Hazardous Waste Generators by
    Industry Type

SIC Code	Manufacturing	No. of generators
34	Fabricated metal products	2,636
28	Chemicals and allied products	24,432
36	Electrical equipment	1,515
33, 35, 37	Other metal products	2,222
20-27, 29-32, 33-39	All other manufacturing	3,208
N/A	Nonmanufacturing	2,074
Total		36,087

Source: McIlvaine [6].

To meet the basic social needs of industrialized, modern
society, a number of industries generate hazardous wastes as
a consequence of the economic process: large chemical and
plastics manufacturers; oil refineries; metals industries;
mining firms; and small businesses such as dry cleaners and
car repair shops. Society demands products to clean clothes
and ovens effectively, to control pests such as fleas and
ticks, and to increase food production through weed control
and growth stimulants (fertilizers). However, many of the
products upon which society has come to rely carry a consequent environmental price for safe and effective disposal.
For example, the production and sales of synthetic organic
pesticides increased from 648 million pounds to 1,189 million
pounds while total sales increased from $262 million to
$4,730 million [17]. These sales occurred to meet agricul-
tural and consumer demand. The quantity of hazardous waste
generated as a function of pesticide manufacture and use also
increased dramatically during this period.

B. Issues and Trends

With over 750,000 businesses generating hazardous waste,
over 10,000 transporters involved in shipping it, and more
than 30,000 storage, treatment, or disposal sites [6], the

magnitude of the problem is readily apparent. The potential
for problems are numerous, ranging from groundwater contamination through leaching to accidents during transportation and
direct contact with hazardous materials. Undetermined amounts
of hazardous wastes have been illegally dumped along roadsides
or into sewers. Of the more than 30,000 known hazardous waste
sites, USEPA has listed 888 of them on its 1986 National
Priorities List that are slated for cleanup under the federal
Superfund legislation [18].

Historically, hazardous waste has been managed and disposed
of by land disposal methods. As awareness of the hazards has
grown, other treatment methods, technologies, and management
systems have been developed and used successfully. The range
of methods fall into three categories: waste reduction,
treatment, and disposal.

1. Hazardous Waste Reduction

Waste reduction, or minimization, has become one preferred
method of hazardous waste management. Industries are using
process modifications, different processes, and recycling to
reclaim and reuse materials. As regulation and disposal costs
continue to rise, industry has an increased incentive to
create minimal wastes and to better manage the wastes it
creates. One study of 33 chemical industries showed that
about half were taking steps to reduce the amount of waste
produced [1]. Waste recycling, through solvent recovery, is
practiced by many industries. Additionally, it is estimated
that 140 facilities in the continental United States perform
solvent recovery on a commercial basis [1]. The increased
availability of this technique is based on economics. Wastes
can also be traded or sold directly to another industry for
use in specific production applications.

2. Hazardous Waste Treatment

Waste treatment is a rapidly growing family of technologies which falls into three general categories: physical,
chemical, and biological. Physical treatments attempt to
isolate or concentrate the waste through techniques such as
evaporation, solidification, or thermoplastic mixing. Chemical treatments change or modify the properties of the wastes;
for example, oxidation-reduction is used to stabilize unstable
chemicals. Incineration, a thermo-chemical process, can be
viewed as chemical treatment. There are approximately 40
facilities offering thermal destruction on a commercial basis
in the United States [1]. Biological treatment uses micro-
organisms to break down organic hazardous wastes. The

use of genetic engineering to assist in this process is
currently underway.

3. Hazardous Waste Disposal

Disposal of hazardous wastes consists of underground
injection and landfilling. Underground injection is suitable
for liquid wastes and is a relatively inexpensive method of
permanent disposal. The liquid waste is injected under
pressure into the earth to depths ranging from several hundred
feet to a mile below the surface.

Land disposal is still the most frequent method of hazardous waste disposal. Approximately 50 commercial land disposal
sites in the United States accept hazardous waste [1]. There
are no such sites in 26 states although a few sites may
reopen after the facilities are upgraded to meet the RCRA land
disposal regulations [1]. Activities in several states indicate that several facilities may be sited in the next few
years. Most incinerators have a substantial link with
landfills.

C. Legislation and Regulations

Hazardous wastes are regulated by a number of federal laws,
including the Resource Conservation and Recovery Act (RCRA),
the Toxic Substances Control Act (TSCA), the Comprehensive
Environmental Response, Compensation, and Liability Act
(CERCLA), and numerous other regulations related specifically
to air or water quality, etc. This section provides a brief
discussion of the more important laws as they affect the
management and disposal of hazardous wastes.

1. Resource Conservation and Recovery Act (RCRA)

The Resource Conservation and Recovery Act of 1976 mandates how hazardous wastes are to be handled and regulated.
Hazardous wastes are subject to a comprehensive set of regulations developed by EPA pursuant to Subtitle C of RCRA. In
addition to the classification of wastes through the application of certain testing criteria, a waste may be deemed hazardous by virtue of EPA listing it on a generic list of hazardous
wastes. In the case of hazardous waste thermal destruction
RCRA specifies the performance standards an incinerator must
meet to be granted a permit to operate.

The incinerator standards require that facilities must
demonstrate that they can achieve the following: (1) 99.99%
destruction and removal efficiency (DRE) for each principal
organic hazardous constituent (POHC); (2) at least 99%

removal of hydrogen chloride from the exhaust gas if stack
emissions are greater than 4 lb/hr (1.8 kg/hr); and (3) particulate emissions no greater than 0.08 g/DSCF (180 mg/m^3)
corrected to 7% oxygen in the stack gas (most state regulations supersede this requirement as they are more stringent)
[16].

2. Toxic Substances Control Act (TSCA)

The Toxic Substances Control Act (TSCA) governs incineration of polychlorinated biphenyls (PCBs). Standards for
incinerating wastes under the Toxic Substances Control Act
are generally more restrictive than those under RCRA. The
TSCA standards related to incineration of hazardous wastes
are as follows: (1) the incinerator must operate at 99.90%
combustion efficiency or better; (2) scrubbers are required to
remove hydrochloric acid from the furnace exit gases; (3) continuous monitoring of furnace temperature, stack oxygen, and
carbon monoxide, and the PCB feed rate; (4) there must be
automatic shutoff of feed supply if monitoring equipment
fails; and (5) incinerators burning solid PCB wastes must
have a 99.9999% destruction and removal efficiency.
   The TSCA regulations go well beyond conventional performance requirements, however. Combustion requirements are
specified by law. The TSCA regulations state that, if a furnace operates at 2200°F (1200°C), the gas will have a residence time in the furnace of at least 2 seconds, and the gas
composition will include at least 3% oxygen (dry basis).
Alternatively, if the furnace operates at 2950°F (1600°C), the
furnace gas residence time is 1.5 seconds and the oxygen
requirement is 1.5%. Some incinerators, notably fluidized
beds, have received TSCA permits operating at 1800°F (982°C)
for 1.3 seconds as is discussed in Chapter VI and VII. These
permits are the exception, however.

3. Comprehensive Environmental Response, Compensation
and Liability Act (CERCLA)

In response to the need to clean up existing contaminated
hazardous waste sites, Congress passed the Comprehensive
Environmental Response, Compensation, and Liability Act
(CERCLA) in 1980. Commonly known as Superfund, this law was
designed to provide money for the cleanup of sites. The law
defines responsibility for cleanup and specifies the
liability.
   Superfund was substantially amended in 1986 by the Superfund Amendments and Reauthorization Act (SARA). This renewal
provides another five years for the program and increases the

budget to over $8.5 billion—more than a five-fold increase
over the initial CERCLA programs. The new legislation has
stricter requirements for remedial actions and will likely
increase the costs of disposal.

D. Issues and Future Trends

Although the basic criteria for determining whether a
waste is or is not hazardous may not change in the near
future, other regulations specifying cleanup standards and the
handling of hazardous wastes remain in flux. A recent
example is the debate over municipal incinerator ash—historically it has been nonhazardous. Now it will likely be cate-
gorized as hazardous and thus must be disposed of accordingly.
As more hazardous waste problem areas are discovered,
additional assistance will be required in the disposal and/or
treatment process. One of the values of incineration is its
ability to not only destroy much of the hazardous materials
by rendering it inert, the process also diminishes the waste
volume significantly.

IV. CONCLUSIONS

The incineration of municipal and hazardous solid wastes
as a treatment technique provides a useful solution, particularly when combined with well-performed landfilling. The
combustion process can handle a great variety of wastes in a
controlled environment with minimal health risks. Public mis-
conceptions and fears currently are resulting in opposition to
incineration as a waste management solution. However, as
landfilling costs rise and recycling is demonstrated to handle
only a portion of the waste stream, municipal and hazardous
waste incineration will provide one reasonable alternative
in appropriate applications.
The remaining chapters of this book provide an examination
of the principles of combustion, and current incineration
processes as well as the associated environmental control
systems.

REFERENCES

1. Baty, C. J., and Perket, C. L. 1984. 1984 Summary Report: Commercial Industrial and Hazardous Waste Facilities. *Pollution Engineering 16*(7):, pp. 24-28, July.

2. Diemer, R. B. et al. 1987. Hazardous Waste Incineration. Encyclopedia of Physical Science and Technology, Vol. 6. Academic Press, San Diego, California.

3. Franklin Associates, Ltd. 1986. Characteristics of Municipal Solid Waste in the United States, 1960 to 2000. Prepared for U. S. Environmental Protection Agency, Office of Solid Waste, July.

4. Glebs, R. T. 1988. Landfill Costs Continue to Rise. *Waste Age 19*(3).

5. Hazardous Waste Management in Washington - Facts and Issues. 1982. The League of Women Voters of Washington, September.

6. McIlvaine, R. W. 1988. A Hazardous Wastes Overview. *Waste Age 19*(6).

7. MITRE Corp. 1984. Profile of Existing Hazardous Waste Incineration Facilities and Manufacturers in the United States. Prepared for Indus-rial Environmental Research Lab, Cincinnati, Ohio, February.

8. National Commission on Materials Policy. 1973. Material Needs and the Environment Today and Tomorrow - Final Report.

9. Nichols, Alan B. 1988. Nation Copes with Garbage's Rising Tide. *Journal WPCR 60*(5).

10. Pavoni, R. et al. 1975. Handbook of Solid Waste Disposal: Materials and Energy Recovery. Van Nostrand, New York.

11. Porter, J. W. 1988. Toward a Recycling Ethic. *Waste Age 19*(3).

12. Radian Corporation. 1983. Characterization of the Municipal Waste Combustion Industry. Final Report. Prepared for the U.S. Environmental Protection Agency, Office of Air Quality Planning and Standards, EPA Contract No. 68-02-3889.

13. Radian Corporation. 1987. Municipal Waste Combustion
     Study: Recycling of Solid Waste. Prepared for the U.S.
     Environmental Protection Agency, Office of Air Quality
     Planning and Standards.

14. Repa, E. W. 1988. A Look at Superfund's Municipal
     Landfills. *Waste Age 19*(5):86-90.

15. Ridzon, G. J., Blum, S. L., White, A. L., and Pearson,
     C. V. 1983. Institutional Issues Concerning Energy From
     Municipal Waste: A Status Report. Argonne National
     Laboratory, Chicago, Illinois.

16. Santolari, J. 1986. Incineration Technology: State-of-
     the-Art Review. Permitting Hazardous Waste Incinerators
     Short Course. USEPA, San Francisco, California.

17. U.S. Bureau of the Census. 1986. Statistical Abstract
     of the United States: 1987 (107th Ed.) U.S. Government
     Printing Office, Washington, D.C.

18. U.S. Environmental Protection Agency. 1986. National
     Priorities List Fact Book. U.S. Government Printing
     Office, Washington, D.C.

Chapter II

FUNDAMENTALS OF MUNICIPAL

SOLID WASTE COMBUSTION

I.   INTRODUCTION

The direct combustion or incineration of municipal solid waste (MSW) is a well proven element of the total waste management system. Facilities have been built at sizes ranging from 10 tons per day (TPD) to 3,000 TPD. Such facilities have several common elements including waste receiving; some level of material processing; a combustion system; most commonly a heat recovery system which includes a boiler, electricity generation system, and/or a heat distribution system; an air quality control system; an ash management system; and the balance of plant as shown in Figure 1. The raw refuse processing system, may include the removal of white goods (refrigerators, stoves, tires, and batteries) by handpicking at mass burn facilities, or to the extent of shredding and trommel separation at a refuse derived fuel (RDF) facility. Once processed, the refuse is burned, useful heat is recovered, and the inert material left on the grate is discharged as bottom ash. Prior to disposal of the bottom ash; ferrous materials may be removed, the ash conditioned, or the ash mixed with flyash. Before the gaseous products of combustion are discharged, they are treated by an air quality control system (AQCS) to reduce emissions of particulates, acid gases, toxic organics, and oxides of nitrogen (NO_x). The important areas which will be treated in this text (Fig. 1) are waste receiving, combustion, and minimization of products of combustion.

The design and operation of the fuel processing, combustion, and air quality control systems require a fundamental understanding of the purposes of MSW combustion, the characteristics of the fuel being burned, and the process of waste combustion. Those fundamentals are review in this chapter.

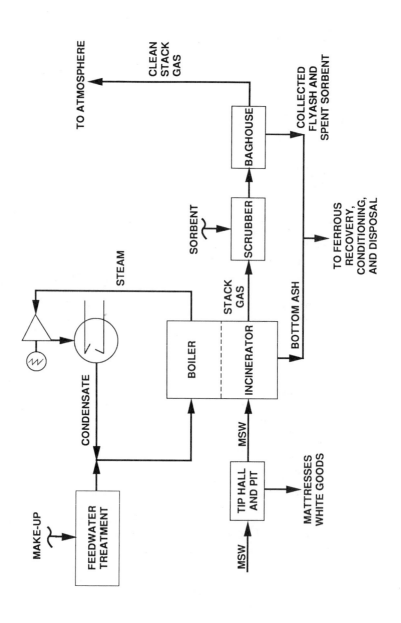

FIGURE 1. Basic components of an MSW mass burn plant.

A. Purposes and Constraints Associated with MSW Combustion

Volume reduction, hence landfill life extension, is the
primary technical purpose of MSW combustion. A typical mass
burn facility can accomplish a 70% mass reduction and a 90%
volume reduction of the incoming municipal waste. A typical
refuse RDF production and combustion facility can achieve
mass reductions of 50-70%, and volume reductions of 75-90%.
These volume and mass reductions reduce the need to site,
acquire, design, install, and operate large land disposal
facilities. Volume reduction is of particular importance in
areas where land is scarce, or where land is unsuitable for
waste disposal. Although, in many cases the ash residue from
incineration cannot be placed into the sanitary landfill. Such
areas would include the corridor from Boston, Massachusetts,
to Washington, D.C.; or areas of Florida where the water table
lies just below the surface of the land. Volume reduction
through incineration must be accomplished in an environmentally sound yet cost effective manner. Consequently, MSW combustion systems typically are constructed and operated with a
heat recovery system. Volume reduction through incineration
is a cornerstone of the waste management process. It comple-
ments other, less ubiquitous techniques, and can be accomplished within an overall framework of waste reduction, recycling, composting, and other voluntary and mandatory strategies. It must be pursued with recognition that land disposal
remains as the ultimate method for managing MSW or the ash
resulting from waste combustion.
Accomplishing volume reduction in a cost effective and
environmentally sensitive manner, and within the framework of
an overall waste management system, requires an understanding
of the fundamentals of waste as a fuel, and of the process of
solid fuels combustion applied to MSW. That understanding is
the focus of this chapter.

B. Overview of the Chapter

The fundamentals of MSW fuels and combustion issues recog-
nize that municipal waste is not a single "fuel." Rather, it
is a heterogeneous mixture of widely dissimilar organic and
inorganic materials which are thrown into a single combustion
chamber. The process of combustion is manipulated through
mechanical design, fuel blending and separation, physical
operation, and accomplished by careful attention to the chemistry of municipal waste oxidation.

This chapter first considers the fundamentals of MSW fuels and combustion by detailing the fuel characteristics of the various materials in the solid waste stream. These properties include proximate and ultimate analyses, calorific values, measures of reactivity, and sources of contaminants such as trace metals and dioxins and furans. The chapter then examines the combustion process in terms of overall mechanisms, determinants of flame temperature, system efficiency, and formation and control of airborne emissions and solid residues. Particular attention is paid to mechanisms associated with formation and control of compounds such as oxides of nitrogen (NO_x), dioxins, and furans. These fuels and combustion evaluations will then lead to conclusions regarding MSW combustion systems.

II. FUEL PROPERTIES OF MSW

The components of MSW include nearly everything discarded by residential, commercial, and light and heavy industrial establishments within any community. As discussed in Chapter I, the composition of waste is a reflection of society: its income level, its need for convenience, its search for cleanliness and sanitation (i.e., the development of disposable diapers), its proliferation of information systems, and its technological advancement (i.e., the development of home computers). The composition of MSW reflects the food production, distribution, and consumption habits of a society. This composition also includes the presence of toxic and/or hazardous substances in the household (i.e., paints and paint thinners, caustic compounds such as oven cleaner, copper cleaner and silver polish, etc.). Ultimately, these wastes of society must be either burned or buried. Consequently, their combustion characteristics as a fuel are of some concern.

A. General Composition of MSW

Combustion is a chemical process that changes the chemical composition of MSW. Because of this process, the chemical composition of MSW must be clearly understood.

1. Waste Product Composition of MSW

Municipal refuse is comprised of many dissimilar materials including various types of paper and plastics, yard trimmings and debris, food waste, wood, textiles, rubber, leather, and

a host of inorganics. Although the percentages of materials
vary somewhat by location and as a function of time, a
"typical" composition of MSW can be estimated as shown in
Table I. Newspaper comprises some 12% of the total waste
stream; magazines and books comprise 3% of the total waste
stream; and other papers, plastics, and packaging materials
when taken together comprise some 31% of the total waste
stream [7]. The need for convenience is demonstrated as our
society shifts from single income households. The amount and
types of convenience packaging in the waste stream is indica-
tive of societal trends.

2. *Variability in the Waste Product Composition of MSW*

Making generalizations about MSW composition is difficult,
however. As noted in Chapter I, MSW composition can vary significantly from community to community, and within any

TABLE I. Estimated Typical Percentage Composition of
Municipal Solid Waste for the Years 1980, 1990, and
2000

Component	Year		
	1980 (%)	1990 (%)	2000 (%)
Paper and paperboard	33.6	38.3	41.0
Yard wastes	18.2	17.0	15.3
Food wastes	9.2	7.7	6.8
Plastics	6.0	8.3	9.8
Wood	3.9	3.7	3.8
Textiles	2.3	2.2	2.2
Rubber and leather	3.3	2.5	2.4
Glass	11.3	8.8	7.6
Metals	10.3	9.4	9.0
Miscellaneous	1.9	2.1	2.1
Total	100.0	100.0	100.0

Source: Seeker, Lanier, and Heap [22].

community depending upon the time of year. Variations in composition resulting from community differences are a function
of community wealth, education level, type of industry, dedi-
cation to such voluntary programs as recycling, and other
similar variables. Studies summarized by the National Center
for Resource Recovery cited by Tillman [27] show food waste
ranging from 9.1% of the total waste stream in Tampa, Florida,
to 16.7% of the total waste stream in Oceanside, New York.
Paper as a percentage of the total waste stream ranged from
21.1% in Flint, Michigan, to 44.6% in Berkeley, California.
Variations in composition, particularly moisture content, can
occur as a function of the waste collection method. Curbside
collection can result in wide moisture content variations of
exposure to weather. This problem is avoided in Sweden, however, since household refuse is picked up from dust bins or
trash rooms within the household. Further, the waste is
"packaged" in large plastic-lined kraft paper sacks. Consequently, the waste remains dry, and its heating value is en-
hanced by the disposable packaging system.

    Variations exist not only with respect to composition,
but also with respect to quantity of waste. As discussed in
Chapter I, the variations in quantity are of considerable
significance in evaluating MSW as a combustible energy
source.

B.   Chemical Composition of MSW

    Combustion is a chemical process where carbon, hydrogen,
sulfur, and fuel-bound nitrogen are oxidized to carbon dioxide, water, sulfur dioxide, and various oxides of nitrogen.
If chlorine is present in the fuel, hydrogen may be prefer-
entially oxidized to hydrogen chloride. The complexities of
MSW combustion are compounded by the fact that municipal
refuse and RDF are not single fuels. Rather, they are mixtures of many dissimilar combustible materials.

1.   Chemical Compounds in Combustible MSW

    The combustible wastes are highly dissimilar and include
various papers, plastics, wood products, textiles, rubber,
food waste, and yard waste. The wide variety of papers include mechanical and thermomechanical pulp based products,
such as newsprint and chemical pulp based products such as
bleached and unbleached kraft paper and board, sulfite papers,
and corrugated boxes. From a simplistic perspective, the
mechanical pulps contain the same constituents as wood while

the chemical pulps largely contain cellulose with varying
quantities of hemicelluloses. For all practical purposes,
chemical pulps are delignified. Although only a minor portion of the waste stream, other papers include those impreg-
nated with asphalt, such as roofing papers, or wax.

Plastics found in the waste stream include nylon, poly-
ethylene films, polystyrenes, polyurethane foams, fiberglass
resins, polyvinyl chloride, and a host of other man-made
polymers. Rubber products are related to plastics, and numerous latex and other rubber compounds also comprise the
waste stream. Finally, wood and wood wastes, cellulosic and
fatty food wastes, lignocellulosic yard wastes, and related
products are part of the waste stream. These components have
varying chemical structures. However, there is generally a
preponderance of aliphatic compounds. It is noteworthy that
wood-based products, such as chemical pulp and paper products
(represented by cellulose) are highly oxygenated. Associated
with these waste components are functional groups, including
not only the hydroxyl groups related to cellulose, and the
methoxyl groups associated with softwood lignins, but also
nitrogen and chlorine functional groups in the plastics
structures. In addition, amine structures are associated with
various plastics, wood wastes, yard wastes, and food wastes;
protein contains 6.25% nitrogen in amine form [6]. These
chemical structures govern the initial combustion/thermal
oxidation reactions associated with MSW incineration.

2. Elemental Analysis and Heating Value of MSW

The ultimate analyses of the various components of MSW
reinforces the concept that MSW is composed of many heterogeneous fuels as shown in Table II. The potential for variation
in the chemical composition of the waste stream then is a
function of a changing waste stream. When a composite proximate (percent moisture, fixed carbon, volatiles, and ash)
and ultimate analysis is constructed for "typical" MSW, the
results are consistently similar (see Table III). Higher
heating value is influenced by waste stream composition as
well, as shown in Table IV. The identifiable combustibles
in the waste stream can range in heat content from 3,265 Btu/
lb (1,814 cal/g) for food waste to 19,860 Btu/lb (11,304 cal/
g) for polypropylene. The difference between food waste and
polypropylene is a six-fold variation in heat content among
combustibles, and illustrates how MSW is a mixture of combustible materials rather than a single fuel.

The chemical composition of MSW, and its components,
indicates not only an aggregation of combustibles with a
modest heating value, but also an assembly of reactive

TABLE II. Composite Analysis of the Basic Components of Municipal Solid Waste

Waste material	Analyses, Wt % as received								
	C	H	O	N	Cl	S	H_2O (%)	Inerts, ash	Btu/lb[a]
Corrug. boxboard	36.79	5.08	35.41	0.11	0.12	0.23	20	2.26	6,233
Newspaper	36.62	4.66	31.76	0.11	0.11	0.19	25	1.55	6,223
Magazines, books	32.93	4.64	32.85	0.11	0.13	0.21	16	13.13	5,446
All other paper	32.41	4.51	29.91	0.31	0.61	0.19	23	9.06	5,481
Plastics	56.43	7.79	8.05	0.85	3.00	0.29	15	8.59	11,586
Rubber, leather	43.09	5.37	11.57	1.34	4.97	1.17	10	22.49	8,433
Wood	41.20	5.03	34.55	0.24	0.09	0.07	16	2.82	6,933
Textiles	37.23	5.02	27.11	3.11	0.27	0.28	25	1.98	6,595
Yard trimmings	23.29	2.93	17.54	0.89	0.13	0.15	45	10.07	4,005
Food waste	17.93	2.55	12.85	1.13	0.38	0.06	60	5.10	3,265
Fines, -1"	15.03	1.91	12.15	0.50	0.36	0.15	25	44.90	2,593

[a] To obtain kcal/g, divide by 1800.

Source: Kaiser [16] as quoted in Ebasco [7].

TABLE III. Typical Proximate and Ultimate Analyses of
Municipal Solid Waste

Analytical parameter	Value (%)
Proximate analysis	
Moisture	25.2
Ash	24.4
Fixed carbon Volatile Matter }	50.4
Ultimate analysis	
Carbon	25.6
Hydrogen	3.4
Oxygen	20.3
Nitrogen	0.5
Chlorine	0.45
Sulfur	0.15
Moisture	25.2
Ash (Inerts)	24.4
Higher heating value	
Btu/lb	4,450
Cal/g	2,472

Source: E. R. Kaiser [16] as reported in Ebasco [7];
also Seeker, Lanier, and Heap [22].

materials. Reactivity can be evaluated by atomic hydrogen/
carbon (H/C) and atomic oxygen/carbon (O/C) ratios, with
higher ratios indicating increased reactivity. Although the
scale is not absolute, the trends exist. Further, the reactivity can be demonstrated by thermogravimetric analysis
(TGA). To demonstrate the relative reactivity, atomic H/C
and O/C ratios for select components of the MSW waste stream
can be compared to H/C and O/C ratios associated with various
ranks of coal as shown in Table V. TGA and DTG curves for

TABLE IV.   Higher Heating Value of Various Components
of the Municipal Solid Waste Stream

Material	Btu/lb	Cal/g
Paper products		
Corrugated boxboard	6,233	3,463
Newsprint	6,223	3,457
Magazines, books	5,546	3,081
Tar paper	11,500	6,389
Waxed paper	11,500	6,389
Plastic coated paper	7,340	4,078
Plastic products		
Nylon	13,620	7,567
Polyethylene film	19,780	10,989
Polypropylene	19,860	11,034
Polystyrene	17,700	9,834
Polyurethane foam	17,580	9,767
Resin bonded fiberglass	19,500	10,834
Biomass type wastes		
Yard waste	4,005	2,225
Food waste	3,265	1,814
Wood waste	6,933	3,852
Other combustibles		
Textiles	6,595	3,664
Ruber and leather	8,433	4,685
"Fines (<1)"	2,593	1,441

Source: Kaiser [16] as reported in Ebasco [7].

various components of the municipal waste stream as shown in
Figs. 2 through 4.   Figure 2 is a TGA curve for cellulose,
representing papers from chemical pulping processes.   Figure 3
is a TGA curve for Red Alder wood, which is used to represent
wood waste, newsprint, and other mechanical pulping products.

TABLE V.   Atomic Hydrogen/Carbon and Oxygen/Carbon
Ratios for Selected Components of the Municipal
Waste Stream

Component	Atomic Ratio	
	H/C	O/C
Corrugated boxboard	1.64	0.72
Newspaper	1.53	0.65
Magazines, books	1.69	0.75
Miscellaneous paper	1.67	0.69
Plastics	1.66	0.11
Rubber, leather	1.50	0.20
Wood	1.47	0.63
Textiles	1.62	0.55
Yard trimmings	1.51	0.56
Food waste	1.71	0.54
Pennsylvania bituminous coal	0.76	0.05
Wyoming subbituminous coal	0.81	0.17
North Dakota lignite	0.81	0.21
No. 2 distillate oil	1.72	N/A
Natural gas (methane)	4.0	N/A

Sources: Kaiser [16] as reported in Ebasco [7]; Babcock
and Wilcox [1].

Figure 4 is a TGA curve for coffee grounds, representing
food wastes. Reactivities for all of these materials are
shown by the rapid rates of weight loss over time once the
onset of weight loss occurs.

Reactivity of plastics has not been shown by TGA analysis.
All plastics, being petrochemical derivatives, are highly
reactive substances. Their reactivity is modified only by
the physical dimensions of the specific product (and the consequent surface/mass ratio).

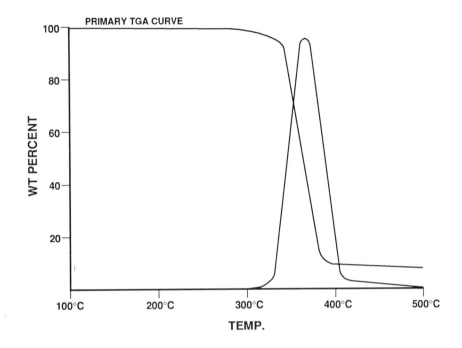

*FIGURE 2. TGA analysis of cellulose (Heating at 15°
C/min). Source: Shafizadeh and DeGroot* [24].

Chemical composition, then, demonstrates that the various
elements of the solid waste stream can be quite dissimilar.
Further, the waste stream when taken as a whole is of modest
heating value. Finally, most combustible components of the
solid waste stream are highly reactive.

3. Chemical Sources of Trace Airborne Emissions

The final issue associated with MSW composition is the
source of potential airborne emissions, including potential
public health problems such as dioxins and furans, and heavy
metals. As the proximate and ultimate analysis demonstrates,
MSW is high in ash content, leading to the generation of particulates which must be controlled. Sulfur and nitrogen contents are low; however, the nitrogen is largely in the food
and yard wastes, is therefore in amine form, and is consequently readily volatilized for subsequent oxidation reactions [29].

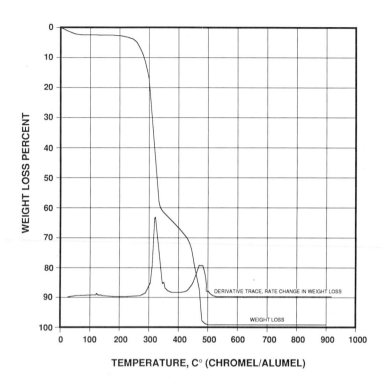

FIGURE 3. TGA curve for Red Alder. Source: Hazen Research [12].

Chlorine is more problematical in MSW, as it contributes to the formation of acid gases (HCl) and chlorinated species of the dioxin and furan classes of compounds. The ultimate analysis shows that various components of MSW, and the total mass of urban waste, can contain appreciable quantities of chlorine. The chlorine in MSW may come from both the paper fraction and the plastics fraction in the waste stream of the community in which it is generated. Heavy metals are contained in all fractions of the waste stream. Research by Haynes et. al [11] focuses on the source of these heavy metals and demonstrates that over 50% of the lead may enter the waste stream in the pigments used to color paper and plastics, and in inks. Pigments, inks, enamels, paints, and lacquers may be responsible for over 50-70% of the chromium, 50% of the cadmium, 60% of the cobalt, 20% of the nickel, and significant portions of the zinc and copper in the waste

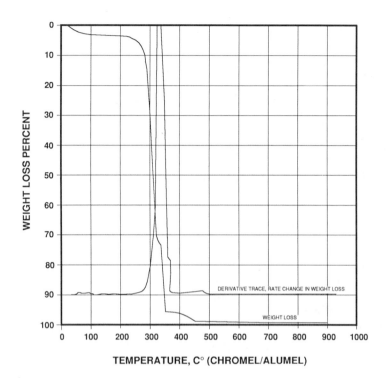

FIGURE 4. TGA curve for coffee grounds. Source: Hazen Research [12].

stream. Research on the composition of waste in Sweden indicates that combustibles may be responsible for less of the
heavy metals than indicated by Haynes, however. Paper and
plastics in the Swedish waste stream contain appreciable but
minor quantities of heavy metals as shown in Table VI. Given
the differences in the time frame of the research, the Swedish results may indicate changes in technology more than
societal (and cultural) waste stream differences.

    Finally, household refuse can contain measurable quanti-
ties of toxic and hazardous wastes, as noted in Chapter I.
These wastes can range from caustics to paint thinners and
solvents, and can include minor quantities of turpentine, oil
based paints, copper cleaner and silver polish, oven cleaner,
and a wide range of other materials. Most communities
attempt to collect these wastes voluntarily, but some

TABLE VI. Percentage Distribution of Metal Content
in Household Solid Waste Fractions

Fraction	Cd	Co	Cr	Cu	Hg	Mn	Ni	Pb	Zn
Plastics	26	1	5	2	10	1	1	5	1
Paper	4	5	7	11	13	18	3	3	11
Animal matter	1	1	1	1	2	1	1	1	1
Vegetable matter	2	1	2	3	6	4	3	2	4
Textiles	1	1	1	2	4	1	1	1	1
Rubber and leather	4	1	42	1	3	1	1	2	9
Metals	60	88	43	22	60	74	87	85	68
Miscellaneous	3	4	3	63	3	4	6	4	6

Source: National Energy Administration and the National
Swedish Environmental Protection Board [26].

residual amounts end up in the mixed municipal waste stream.
Improperly treated, these wastes can result in subsequent
problems.

C. Conclusions Regarding Municipal Waste as Fuel

The previous analysis highlights the perspective that MSW
is not a single fuel, but a mixture of many combustible fuels.
These fuels vary, and exist in the waste stream for social and
economic reasons. Although heterogeneous, the fuels must be
burned in a single mass.

The waste stream contains materials that have vastly different chemical compositions, heating contents, reactivities,
and contributions to the formation of airborne emissions including dioxins, furans, and heavy metals. Taken together,
however, these materials yield a composite with properties
similar to many biomass fuels. Consequently, the process of
MSW incineration can be readily understood through analysis of
wood and biomass combustion.

III. THE PROCESS OF MSW COMBUSTION

The compositional data presented above provide a basis for
understanding the process of MSW combustion which is being
used in mass burn and RDF systems throughout the United States,
Europe, and Japan. The manipulation of the process provides
for maximum heat release, maximum volume and mass reduction
in the materials going to landfill, and the minimum formation
of undesirable airborne emissions.
MSW is a carbonaceous substance containing significant
quantities of hydrogen, sulfur, and chlorine. The general
summary equation for hydrocarbon combustion stoichiometry,
assuming oxygen as the oxidant, is as follows:

$$C_v H_w + (v + w/4)O_2 => vCO_2 + \frac{w}{2} H_2O \qquad (2-1)$$

In this equation, lower case letters represent quantities of
the chemical elements: carbon, hydrogen, and oxygen. This
equation is expanded as follows for chlorinated hydrocarbons:

$$C_v H_w Cl_x + (v + w/4 - x/4)O_2$$
$$=> vCO_2 + XHCl + (w-x)/2H_2O \qquad (2-2)$$

For a hydrocarbon type material containing both chlorine and
oxygen, the summary stoichiometric equation is as follows:

$$C_v H_w Cl_x O_y + (v + w/4 - x/4 - y/2)O_2$$
$$=> vCO_2 + XHCl + (w-x)2H_2O \qquad (2-3)$$

If air is the oxidant, then the term $(v + w/4 - x/4 - y/2)$ is
multiplied by 3.76. If sulfur is present, it is handled in a
manner analogous to carbon. The above equations summarize
the stoichiometry for the oxidation of any material bearing
carbon, hydrogen, chlorine, and oxygen. They are the foundation of expressing any combustion process. At the same time,
however, equations 2-1 to 2-3 mask the highly complex mechanisms associated with solid fuels combustion. And it is the
manipulation of solid fuel combustion mechanisms that permits
sound design and operation of any MSW burner.

A. A Review of the Solid Fuels Combustion Mechanism Applied
    to MSW

    Combustion, or hydrocarbon oxidation, is a complex chemi-
cal process based upon free radical reaction sequences. One
overall schematic for such reaction sequences is shown in
Fig. 5. As discussed previously MSW contains highly volatile,
reactive hydrocarbon solids. In any incineration environment,
these solids undergo the following overall reaction sequences:
(1) particle drying and heating to reaction temperature;
(2) pyrolysis, sublimation, or other solids volatilization
reactions; (3) gas-phase pyrolysis and free radical generation (chain initiation); (4) primary oxidation reactions

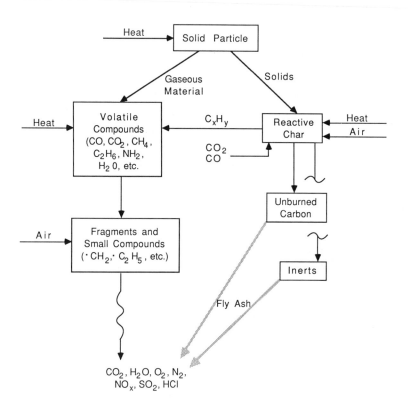

FIGURE 5. Overall MSW combustion mechanism.

(chain propagation reactions); (5) post-combustion (chain termination) gas phase reactions; and (6) gas-solids char oxidation reactions. These overall reaction sequences include numerous highly important specific mechanisms, and each is reviewed below.

1. *Drying and Heating the Solids to Reaction Temperature*

All combustion of solid materials begins by the physical process of heating the mass of fuel or waste. This heating drives off any moisture contained in the material. Further it raises the temperature of the fuel or waste to a level where the chemical processes associated with thermal oxidation commence. The critical variable associated with this phase of the overall mechanism is the temperature of reaction for the solids. Approximate temperatures of reaction initiation can be estimated from TGA curves. For chemical pulp based papers, such temperatures of reaction onset are about 570°F (300°C). For newsprint and wood waste, temperatures of reaction onset may be about 480°F (250°C). For food wastes such as fruit (i.e., grape pomace) or coffee grounds, the temperatures of onset may be 400-450°F (200-230°C)[12,24]. As a practical matter, the ignition temperature of paper and plastics is somewhat higher than the temperature of reaction (pyrolysis) onset, however. It is important to note that the temperatures at which reactions begin are relatively low for the components, and mass, of MSW.

Heating and drying are endothermic processes. Consequently energy consumption and rates of reaction are additional important variables associated with this phase of the combustion process. Energy consumption is a function of the moisture content, heat capacity, and thermal conductivity of the mass of heterogeneous material fed to the combustor. Heating rates are a function of particle size, particularly with respect to the minor dimension as well as the effective depth of material on the grate. Heating rates also are a function of the thermal conductivity of the particle, the thermal head or temperature in the combustor, and the radiative heat transfer mechanism. Finally, heating rates of the combustibles in MSW are a function of the extent and nature of the noncombustible materials in the feed to the incinerator. Increasing the ferrous content of the waste being fed to a mass burn unit from 5% to 15% of the total mass will substantially reduce the availability of energy to the combustible fraction for heating and drying sequences. Energy consumption and reaction rates vary substantially as a function of the specific MSW composition at any given location or moment in time.

2. Volatilization Mechanisms

Once the temperature required for reactive onset is
reached, the volatilization mechanisms can proceed. Pyrolysis
is the dominant mechanism for the production of volatiles from
MSW and its component feedstocks. Pyrolysis is the heating of
any organic substance in the absence of an oxidant. Once a
portion of the fuel particle reaches reaction temperature,
pyrolysis commences. Typically, pyrolysis of an oxygenated
hydrocarbon solid is summarized as follows:

$$\text{Solid} => CO_2 + CO + H_2O + CH_4 + C_xH_y + NH_z \qquad (2\text{-}4)$$

$$+ \text{ other noncondensible volatiles} + \text{tar} + \text{char}$$

The tar tends to be a high molecular weight condensed product
containing polycyclic aromatic hydrocarbons. The char is not
pure carbon as it also contains some hydrogen and oxygen.
Bradbury and Shafizadeh [3] found that chars from the pyrolysis of cellulosic materials typically have compositions with
the empirical formula of $C_{6.7}H_{3.3}O$.

Pyrolysis mechanisms are complex, and can be manipulated
to produce proportionately more volatiles or more char as
illustrated by the cellulose degradation mechanism shown in
Fig. 6. This mechanism is relevant to all paper products and
many food, yard, and wood wastes in MSW. The process begins
by cleavage of the glucosidic linkage. With high temperatures, the pyrolysis reaction sequence leads to ring opening
and fragmentation of the glucose linkage. With low temperatures, reactions include dehydration and condensation leading
to char formation. The mechanism in Fig. 6 illustrates the
relative distribution of pyrolysis products between gaseous
compounds and char as a function of fuel particle size, moisture content, and reaction temperature. Small particles of a
given feedstock (i.e., writing papers) if dry and pyrolyzed
rapidly, will produce a higher percentage of volatiles than
large particles heated more slowly. For all components of
MSW, however, the dominant pyrolysis product is volatiles, as
shown previously with the TGA curve presented in Fig. 3. It
is also a consequence of the high atomic H/C and O/C ratios
associated with the components of MSW (Table V). As pure
compounds, most organic constituents in MSW will pyrolyze to
75-90% volatiles (wt basis) and 10-25% char. Under combustion conditions, with the presence of significant quantities
of moisture and inorganics in the waste stream, a more likely
distribution is on the order of 60-70% volatiles and 30-40%
char. This distribution can be manipulated by combustion
conditions.

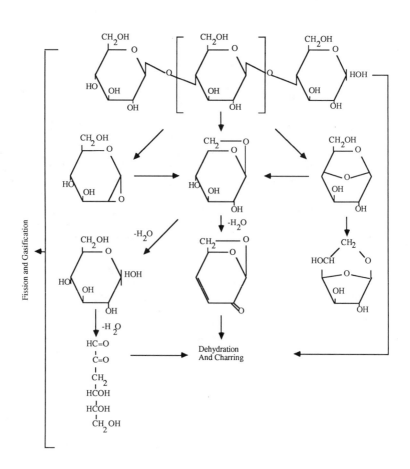

FIGURE 6. Mechanism for pyrolysis of cellulose.
Source: Shafizadeh and DeGroot [23].

The volatiles produced have been identified in equation
2-4 as CO, CO_2, H_2O, CH_4, and other light hydrocarbons.
Additional volatiles include amine species and halocarbons,
depending upon the specific fraction of the waste stream
being pyrolyzed. Volatiles produced include not only stable
gaseous compounds but also radical fragments. These fragments
may include such species as hydroxy radicals, methoxyl radicals, and other pyrolysis products resulting from the cleavage
of functional groups from the main skeleton of any given compound (i.e., cellulose, lignin, styrene) as shown previously
in Fig. 2. Similarly, the chars produced by pyrolysis contain numerous radical sites available for subsequent reaction [3]. Amines and halocarbons are potential sources for
pollutants. Radical fragments and reactive sites are part of
the chain initiation process.

3. Gas Phase Pyrolysis and Free Radical Generation

Once volatilization has proceeded, the gaseous compounds
undergo pyrolysis and fragmentation to produce the free radicals essential in the combustion process. A near-infinite
number of reaction sequences can be postulated for the generation of radicals and fragments [8, 9, 19, 28, 31]. Representative reaction sequences are as follows:

$$\text{solids} \Rightarrow CH_3COOH + \text{other products (see Rxn 2-4)} \qquad (2\text{-}5)$$

$$\text{solids} \Rightarrow CH_3CHO + \text{other products (see Rxn 2-4)} \qquad (2\text{-}6)$$

$$CH_3COOH \Rightarrow CH_4 + CO_2 \qquad (2\text{-}7)$$

$$CH_3CHO \Rightarrow CH_4 + CO \qquad (2\text{-}8)$$

$$C_2H_6 + M \Rightarrow 2CH_3 + M \qquad (2\text{-}9)$$

$$CH_3 + C_2H_6 \Rightarrow CH_4 + C_2H_5 \qquad (2\text{-}10)$$

$$C_2H_5 + M \Rightarrow C_2H_4 + H + M \qquad (2\text{-}11)$$

$$H + C_2H_6 \Rightarrow H_2 + C_2H_5 \qquad (2\text{-}12)$$

$$CH_4 + M \Rightarrow CH_3 + H + M \qquad (2\text{-}13)$$

In all of these reactions, M denotes any heat removing species.
The wide diversity of available pyrolysis products makes precise postulation of chain initiation reactions impossible.
However, two important principles are as follows: (1) a
diverse number of radical generation or chain initiation

pathways exists; and (2) gas phase pyrolysis of organic products, not the cleavage of the diatomic oxygen molecules in the combustion air, provides for chain initiation and the start of the combustion process.

4. Gas Phase Combustion Reactions

Once chain initiation has commenced, the primary gas phase (flaming) combustion reactions can proceed. Again, the list of probable and possible reactions are difficult to predict. One of the very many possible sequences is as follows:

$$CH_3 + O_2 + M \Rightarrow CH_3O_2 + M \qquad (2\text{-}14)$$

$$CH_3O_2 \Rightarrow CH_2O + OH \qquad (2\text{-}15)$$

$$CH_2O + OH \Rightarrow CHO + H_2O \qquad (2\text{-}16)$$

The importance of the reaction sequence shown above is only that it indicates the generation and use of hydroxy radicals. Hydroxy radicals are the most reactive species in the combustion system.

The primary gas phase combustion reactions are followed by post-flame or secondary combustion reactions, including the following representative sequence:

$$CHO + OH \Rightarrow CO_2 + H_2 \qquad (2\text{-}17)$$

or,

$$CHO + OH \Rightarrow CO + H_2O \qquad (2\text{-}18)$$

$$CO + O_2 \Rightarrow CO_2 + O \qquad (2\text{-}19)$$

$$CO + O \Rightarrow CO_2 \qquad (2\text{-}20)$$

$$H + OH \Rightarrow H_2O \qquad (2\text{-}21)$$

Post combustion reactions are responsible for completing the process of solid waste oxidation. These reactions include the sequences required for oxidation of carbon monoxide, where rates of oxidation are slow. Further, these reactions reduce the presence of nonmethane hydrocarbons and other incomplete products of combustion in the stack gas.

5. Char Oxidation (Gas-Solids) Reactions

The final segment of the MSW combustion process to be considered is char oxidation. Typically, char oxidation supplies
the heat for fuel drying and heatup, and solids pyrolysis, in
the MSW combustor. Further, char oxidation reactions contribute to the overall incineration process.

Cellulosic chars are porous, and contain both hydrogen and
oxygen as discussed previously. Further, such chars contain
numerous radical sites [3]. Several mechanisms have been
postulated for char oxidation including chemisorption of
oxygen on the surface of the char [3], leading to the following reactions:

$$C + O_2 \Rightarrow C(O) \Rightarrow C(O)_m \Rightarrow CO + CO_2 \qquad (2\text{-}22)$$

$$C + O_2 \Rightarrow C(O)_x \Rightarrow CO_2 \qquad (2\text{-}23)$$

Reaction 2-22 occurs rapidly due to the mobile $C(O)$ site as
designated by the subscript m, while reaction 2-23 occurs more
slowly due to the stability of the $C(O)$ site.

Mulcahy and Young [17] proposed a second mechanism for
char oxidation from cellulosic materials as follows:

$$2OH + C \Rightarrow CO + H_2O \qquad (2\text{-}24)$$

$$OH + CO \Rightarrow CO_2 + H \qquad (2\text{-}25)$$

In this mechanism, hydroxy radicals may be generated by
pyrolytic cleavage of hydroxy functional groups contained
within the mass of fuel (i.e., OH functionalities from the
cellulose or paper fraction). An alternative source of
hydroxy radicals is dissociation of moisture found in the
fuel. The dissociation of the water molecule is a less
likely source of OH radicals due to the energy required for
bond dissociation. However, this source is not totally
improbable since there is an abundance of heat. Once the
hydroxy radicals are generated, the char oxidation proceeds
rapidly.

The final mechanisms proposed for char oxidation are gasification reactions such as the Boudouard and steam-carbon
reactions shown below:

$$C + CO_2 \Rightarrow 2CO \qquad (2\text{-}26)$$

$$C + H_2O \Rightarrow CO + H_2 \qquad (2\text{-}27)$$

Both reactions are favored at the high temperatures associated with char oxidation that have been measured in existing
MSW combustors (see Table VII). Because most of the moisture
will have left the fuel prior to gas-solids oxidation reactions, the reaction 2-26 is more likely to dominate than
reaction 2-27.

6. Summary

The general mechanisms associated with MSW combustion are
highly complex due to the heterogeneous collection of combustible materials in the waste stream. The complexity of the
reaction sequence provides the opportunities required for
manipulation of the combustion process for optimal system
performance.

B. Considerations of MSW Combustion Temperature
    and Efficiency

Thermal oxidation is a high temperature reaction generating heat which may be put to useful purpose (i.e., generating
steam used for electricity production). Temperature and heat
release are critical to achieve maximum system efficiency and
cost control. Temperature manipulation is essential for controlling airborne emission formation. This section, therefore,

TABLE VII. Tabulated Equilibrium Constants for
     Gasification Reactions (Values are Dimensionless)

T (°F)	T (°C)	$C+H_2O \rightleftharpoons CO+H_2$ K_1	$C+CO_2 \rightleftharpoons 2CO$ K_2
1100	593	0.219	0.0846
1200	649	0.675	0.360
1300	704	1.83	1.30
1400	760	4.47	4.07
1500	816	9.95	11.4
1600	871	20.5	28.8
1700	927	39.5	66.8

focuses on flame temperature, flame intensity and rate of combustion, and combustion system (hardware) thermal efficiency.

1. Flame Temperatures Associated with MSW Combustion

Flame temperature is a critical parameter essential in the manipulation of the combustion process. It can be calculated by iterative solution of the following equation [13]:

$$\sum H_p = \sum n_i \int_{298}^{T} c_{p_i}\, dT + \sum n_i \lambda_i \qquad (2\text{-}28)$$

Where H_p is enthalpy of the products (cal/g-mole), T is flame temperature (°K), c_{p_i} is heat capacity, n is g-moles, and λ is latent heats associated with phase changes (i.e., heat of vaporization).

With the advent of computer technology, the National Aeronautics and Space Agency (NASA) developed a program for flame temperature calculation based upon the JANAF Thermodynamic tables [10]. Extensive use of this model demonstrates that flame temperature curves are parabolic with respect to equivalence ratio, or excess oxygen (Fig. 7). Equivalence ratio is calculated on a molar basis as follows:

Equivalence (ϕ) = (fuel/air, actual)/(fuel/air,

stoichiometric)   (2-29)

Within the region where the equivalence ratio is 0.4 - 0.9 (excess air is 150% to 11%), adiabatic flame temperatures are essentially linear [1,30]. Consequently, linear regression equations can be used to approximate adiabatic flame temperature.

Adiabatic flame temperature regression equations have been developed for the combustion of MSW. The equation using English units is shown below:

$$T_f = 0.108(HHV) + 3476(\phi) - 4.544(M) + 0.59(T_a - 77)$$
$$- 287 \qquad (2\text{-}30)$$

Where T_f is adiabatic flame temperature in °F, HHV is higher heating value of the fuel (Btu/lb), M is the as-received moisture percentage of the fuel, and T_a is the temperature of the air as it enters the combustor (after the air

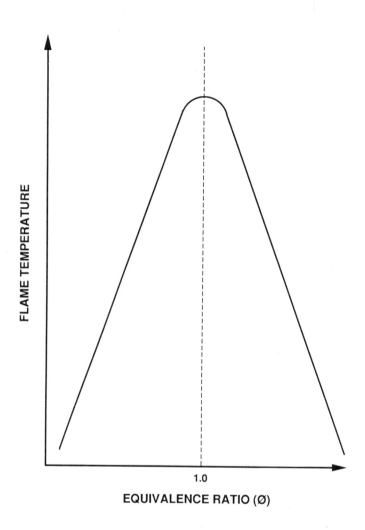

Figure 7. Flame temperature as a function of equivalence ratio.

preheater). Equivalence ratio is used to express the quantity of air used. The relationship between equivalence ratio and excess air can be calculated using the following equation:

$$\phi = 0.7242 - 0.0022 \times EA \text{ percent} \tag{2-31}$$

Equation 2-31 can then be substituted into equation 2-30 to calculate flame temperature on an excess air basis rather than an equivalence ratio basis. Alternatively, equivalence ratio can be calculated from oxygen in the dry stack gas using the following equation (for excess air) presented in Steam [1]:

$$\text{EA percent} = 100 \times (O_2 - CO/2)/[0.264N_2 - (O_2 - CO/2)]$$

$$(2/32)$$

The use of these equations yields a family of curves for MSW combustion as shown in Fig. 8. Flame temperatures will be in the range of 1800°F to 2300°F (980-1260°C) under most operating conditions.

2. Reaction Rates Associated with MSW Combustion

Reaction rates can be understood in terms of flame intensity, a phenomenon closely related to flame temperature. Flame intensity can be evaluated using the equation developed by Shafizadeh and DeGroot [24]:

$$I = dw/dt \ (h) \qquad\qquad (2-33)$$

Where I is intensity, dw/dt is the rate of weight loss as measured by taking the first derivative of the TGA curve as shown in Figs. 3 through 5, and h is the heat of combustion of the fuel. The reactivity of MSW constituents when measured by rate of weight loss, as discussed previously, illustrates that solid waste is very volatile, and the rates of combustion will be correspondingly high. The rate of weight loss for any fuel constituent is governed by inherent volatility, fuel particle size with respect to minor dimension, and moisture content. The impact of these variables has been shown by Junge[14] for wood wastes and, by inference, for the paper and yard waste constituents of the solid waste stream. The data developed by Junge demonstrate that, as particle sizes are reduced from 4" to 1/32" increasing the surface-to-mass ratio by an order of magnitude (a factor of 10), and as moisture content is reduced from 45% to 5%, combustion rates can increase by two orders of magnitude (a factor of 100).

Heat of combustion of various solids as shown in Table IV is the final parameter in determining flame intensity. Data illustrate the wide range of materials in the solid waste stream, some with heats of combustion equalling petroleum fuels while others with modest heats of combustion. Consequently, the rates of combustion will vary widely for different components of the solid waste stream.

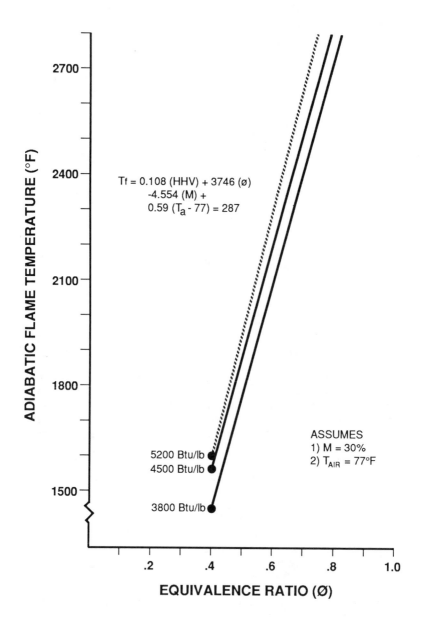

FIGURE 8. Calculated adiabatic flame temperature for MSW combustion.

3. Efficiency of MSW Combustion

Flame temperature and intensity are, indirectly, determinants of MSW combustion system thermal efficiency. Thermal
efficiency can be calculated as follows:

$$n = [1-(HL/HI)] \times 100 \qquad\qquad (2-34)$$

Where n is efficiency (percent), HL is the sum of all heat
losses, and HI is the sume of all heat inputs. HL desig-
nates such heat losses as dry stack gas losses, moisture
losses, unburned carbon, inert solids losses, radiation, and
manufacturers margin.
Efficiency can also be viewed in terms of exergy. Exergy
is defined mathematically, according to the Kalina cycle, as
follows:

$$Exergy = H [1-(T_0(\ln T_1/T_2)/(T_1-T_2))] \qquad\qquad (2/35)$$

Where E is exergy, or the availability of energy from a given
source to perform useful work; H is the initial enthalpy of
the source (i.e., MSW) as evaluated between T_1 and T_2; T_1 is
the initial (hot) source temperature or flame temperature; T_2
is the final (stack) temperature; and T_0 is the ambient temperature.
Flame temperature, combustion rate, and system efficiency
are all interrelated phenomena. These variables can be manipulated by design and operational control, using both the
combustion process per se, and the combustion (hardware)
system.

C. Formation and Control of Airborne Emissions and Solid Residues from MSW Combustion

The final fundamental consideration associated with MSW
combustion is the generation and control of airborne emissions, including particulates, acid gases (sulfur dioxide, SO_2
and hydrogen chloride, HCl), oxides of nitrogen (NO_x), carbon
monoxide (CO) and hydrocarbons (HCs), and dioxins (PCDD) and
furans (PCDF). Particulate emissions are governed by the ash
and inert content of the fuel, the firebox configuration, and
combustion conditions. They are influenced by manipulation
of the combustion process to the extent that emissions can be
altered by the residence time of solids on the grate system,
by the percentage of excess air used, and by the underfire/
overfire air ratio [15]. Acid gas emissions are determined
largely by the sulfur and halogen content of the fuel. They

can be influenced to a minor extent by firing methods [5] but are not particularly influenced by manipulation of the combustion process.

Significant emissions which can be influenced by manipulating the combustion process include NO_x, CO and HCs, and dioxins and furans. These emissions can be affected not only by configuration of the firebox and the distribution of air, but also by complete manipulation of the combustion chemistry. Consequently, NO_x emissions and organic materials emissions are reviewed in more detail below.

1. Emissions of Oxides of Nitrogen

There are two sources of NO_x from MSW combustion: thermal NO_x and fuel NO_x. The relative importance of each source varies depending upon fuel composition and combustion conditions.

a. Thermal NO_x Formation. Thermal NO_x results from the dissociation of diatomic nitrogen in air, and subsequent oxidation of the nitrogen atoms. The dominant mechanics for thermal NO_x formation is the Zeldovich mechanism shown below (a more complete discussion of thermal NO_x formation and the Zeldovich mechanism can be found in Glassman [9]; Palmer [19]; or Edwards [8]:

$$O + N_2 \Rightarrow NO + N \qquad\qquad (2-36)$$

$$N + O_2 \Rightarrow NO + O \qquad\qquad (2-37)$$

Both reactions 2-36 and 2-37 are reversible, high temperature reactions with the following forward rate constants [19]:

$$k_{(2-36)} = 1.36 \times 10^{14} \exp(-75.4\text{kcal}/RT) \text{ cm}^3 \text{ mole sec}$$
$$(2-38)$$

$$k_{(2-37)} = 6.4 \times 10^9 T \exp(-6.25\text{kcal}/RT) \text{ cm}^3 \text{ mole sec}$$
$$(2-39)$$

Reaction 2-36 not only has a very high activation energy, but is also controlled by the availability of oxygen atoms. Consequently, lower temperatures, or low levels of excess air, can restrict the formation of thermal NO_x.

Once formed, NO_x molecules may be "frozen" by reducing temperature. The following rate constant equation for the reverse of reaction 2-37 illustrates this point [19].

$$k_{(-2-37)} = 1.55 \times 10^9 T \exp(-38.6 kcal/RT) \text{ cm}^3/\text{mole sec}$$

$$(2-40)$$

While these are the dominant thermal NO_x formation reactions, others may make contributions including [19]:

$$N_2 + O_2 \Rightarrow N_2O + O \qquad\qquad (2-41)$$

$$N_2O + O \Rightarrow 2NO \qquad\qquad (2-42)$$

$$N + OH \Rightarrow NO + H \qquad\qquad (2-43)$$

$$N_2 + OH \Rightarrow N_2O + H \qquad\qquad (2-44)$$

and the "prompt NO" formation mechanism in fuel rich regions or flames, identified by the following reaction eluciated by Fenimore [19]:

$$N_2 + CH \Rightarrow CHN + N \qquad\qquad (2-45)$$

Thermal NO_x is formed only at very high temperatures. Pershing and Wendt [20] demonstrated that actual flame temperatures must be in the region of 2700°F (1480°C) and adiabatic flame temperatures have to exceed 3200°F (1760°C) for thermal NO_x emissions to be significant. Such flame temperatures are not common in municipal waste incinerators, although they may be found in localized "hot spots" where high concentrations of plastics exist momentarily. Further, they could be experienced in some RDF combustors using relatively low levels of excess air.

  b. *Fuel NO$_x$ Formation.* Fuel NO_x results from the oxidation of nitrogen contained in the fuel. Research to date shows that the formation of the fuel NO_x is largely independent of temperature [20, 21, 28].
     As waste is combusted, nitrogen typically leaves the fuel as a volatile, commonly as an amine radical. This volatilization is independent of temperature; further, once pyrolysis has commenced to a vigorous stage, nitrogen volatiles will be evolved preferentially to carbon volatiles [29]. If sufficient oxygen is present the amine radicals will be oxidized to NO. Alternatively, in an oxygen deficient environment, the nitrogen from the fuel will become diatomic nitrogen. This use of an oxygen deficient environment is the basis for staged combustion (see Munro [18] for a complete discussion of staged combustion manipulation). An overall schematic for fuel nitrogen reactions are shown in Fig. 9.

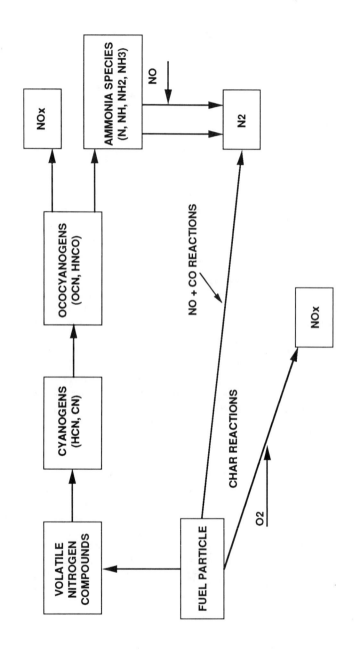

FIGURE 9. Fuel bound nitrogen reaction pathways.

Formation of NO_x from fuel bound nitrogen is the dominant mechanism for creation of this airborne emission from MSW incineration. Nitrogen sources include yard wastes, food wastes, some plastics, and a host of miscellaneous sources. Typically, the nitrogen in MSW is reactive and readily volatilized. Further, as can be calculated from equation 2-30, the temperature regime associated with MSW combustion is more conducive to fuel NO_x formation than thermal NO_x formation.

2. Formation and Control of Organic Emissions

Organic emissions include CO, HCs, and dioxins and furans. They are analyzed here as a single unit because the conditions which favor or minimize formation of all organic compounds are identical as shown by Seeker, Lanier, and Heap [22].

a. Carbon Monoxide and Hydrocarbons. Carbon monoxide and hydrocarbons are emitted when insufficient oxygen exists for complete combustion, or when excess air levels are so great that flame temperatures are reduced severely. EER Corporation [22] developed a curve as shown in Fig. 10 depicting CO emissions from MSW combustion as a function of excess air. Hydrocarbon emissions mirror CO emissions with respect to excess air. Based upon these data, it is apparent that manipulation of the combustion process can be used to control CO and HCs.

b. Dioxins and Furans. Municipal waste incinerators have been found to be sources of dioxins and furans. As discussed in Chapter I, dioxins and furans are families of compounds. The species 2,3,7,8-tetrachlorinated dibenzo-dioxin and the related 2,3,7,8-tetrachlorinated dibenzo-furan are considered the most toxic of these compounds.

Four general mechanisms have been postulated to explain dioxin formation from MSW incineration [4, 22, 25]:

1. the presence of dioxins in the incoming waste, and the passage of those dioxins through the incinerator in unreacted form;
2. the formation of dioxins and furans from related chlorinated precursors;
3. the formation of dioxins and furans from appropriate organic species (i.e., building blocks of lignin) and a chlorine donor; and
4. the formation of dioxins and furans on flyash particles from the condensation of organics and chlorine atoms on the surface of the solid particle.

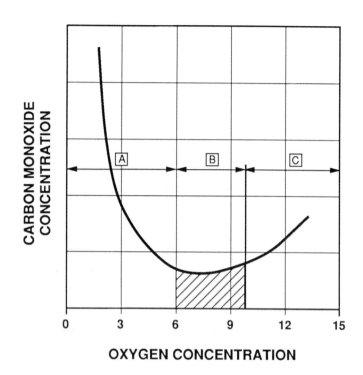

FIGURE 10. Excess oxygen (excess air) and carbon monoxide emissions, showing the appropriate range for MSW
incinerators. Note that Region A involves insufficient
oxygen for complete combustion of the carbon in the
waste while Region C involves "cold burning," and
unfavorable conditions for oxidizing the CO formed.
Region B is the appropriate operating region.
Source: Seeker, Lanier, and Heap [22].

Of these, the second and third mechanisms as shown in Fig. 11
are considered among the most likely as sources of dioxins
and furans. The third mechanism is fundamental to the controversial paper, "Trace Chemistries of Fire," authored by
Bumb et al [4].

Research by Barton et al [2] includes summarizing the
numerous previously proposed mechanisms for dioxin and furan
emission formation, and then evaluating analytical data
resulting from testing the incinerator in Quebec City. That
research indicates that (1) dioxin formation involves

Formation from Related Chlorinated Precursors

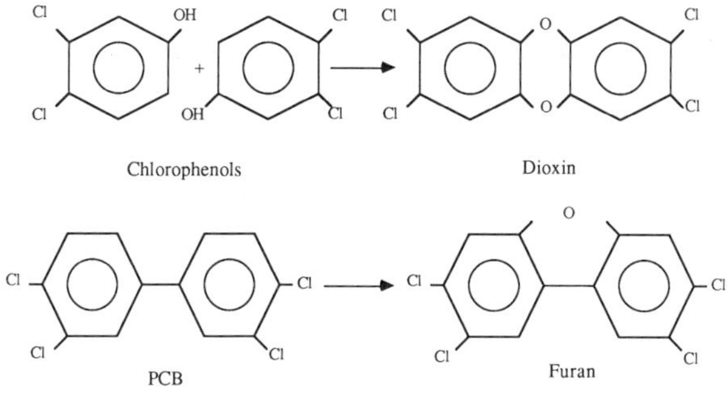

Chlorophenols   Dioxin

PCB   Furan

Evidence: PCDD/PCDF on soot form PCB tires
            Lab and bench studies of PCB, Chlorinated Benzene,
            and Chlorinated Phenols yielded PCDD/PCDF

Formation from Organics and Chlorine Donor

PVC   ⎫
       ⎬ + Chlorine donor   ──────▶ PCDD/PCDF
Lignin ⎭   NaCl, HCl, Cl$_2$

Evidence: Lab scale tests of vegetable matter, wood, lignin,
            coal with chlorine source yielded PCDD/PCDF

FIGURE 11. Dominant combustion mechanisms for dioxin and furan formation. Source: Seeker et al. [22].

homogeneous gas-phase reactions between hydrocarbons and chlorine donors, and (2) that dioxin and furan formation is substantially promoted by the presence of flyash particles. The research by Barton et al. [2] demonstrated that there is sufficient excess chlorine from many sources in the waste to facilitate this mechanism. Further, the research demonstrates that the flyash particles appear to act as catalysts. The research by Barton et al. proposes that the trace quantities of hydrocarbons which escape complete oxidation to CO_2 and H_2O in the furnace form dioxins and furans on the surface of entrained flyash particles downstream of the combustion zone (i.e., in the economizer) at temperatures of 480-660°F (250-350°C). These compounds, with inorganic chlorine, form

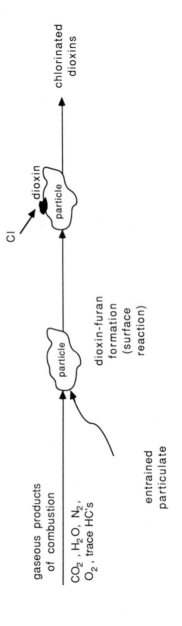

FIGURE 12. Speculative mechanism for formation of chlorinated dioxins and furans from MSW. Source: Barton, Clark, Lanier, and Seeker [2].

2,3,7,8 TCDD and 2,3,7,8 TCDF, and other chlorinated dioxin
and furan compounds. This mechanism is summarized in Fig. 12.
   Dioxin formation and destruction are highly sensitive to
temperature and oxygen availability. Control of dioxin and
furan formation is generally considered to require temperatures exceeding 1800°F (980°C) for residence times in excess
of 1 sec. Further, oxygen availability at the grate is
required to exceed the stoichiometric volume of O_2 required
for complete oxidation, such that reducing conditions are
avoided (based upon calculations from Table 8-1 of Seeker,
Lanier, and Heap [22]. Excess air levels, at a minimum, must
exceed 50-60% [22]. Further, it is generally shown that increasing temperatures fragments the potential dioxin precursors, making dioxin formation in the post combustion zones of
the system less likely.

IV. PRINCIPLES OF MSW COMBUSTION: CONSEQUENCES FOR
     SYSTEM DESIGN AND OPERATION

   MSW combustion is a complex chemical reaction system which
involes subjecting a myriad of combustible materials to processes of heating and drying, pyrolysis and volatilization,
gas phase pyrolysis and oxidation reactions, and gas-solids
(char oxidation) reactions. The issues associated with solid
waste combustion include manipulation of the mechanism
sequences for complete combustion and volume minimization,
maximum energy generation and recovery for cost effectiveness,
maximum temperature generation for dioxin destruction, and
combustion control for minimization of all airborne emissions.
   Methods for manipulating the combustion process begin with
the fuel itself. The fuel may or may not be prepared in order
to remove some pollutants (i.e., heavy metals, fuel bound
nitrogen), reduce the inert loading on the system, and promote
more uniformity in the fuel. Manipulation of the fuel
resource ranges from encouraging voluntary recycling and composting through the production of RDF products at various
degrees of refinement.
   Manipulation of the combustion mechanism then involves
combustion system design and operation--utilizing the 3 "Ts"
of combustion: time, temperature, and turbulence or mixing.
Ensuring adequate oxygen to all portions of the combustion
system, and adequate mixing, is essential. Manipulating temperature and residence time for CO, hydrocarbon, and dioxin
control without promoting the formation of thermal NO_x is a
second issue. Manipulation of flame and final stack temperatures for maximum efficiency is also essential.

Numerous systems provide combustion of MSW, including a variety of mass burn incinerators (Chapter III). Such incinerators include conventional waterwall designs and rotary kilns. Other methods for manipulating MSW combustion involve upgrading the fuel through RDF production processes, and then combusting that RDF in spreader-stoker systems, fluidized bed boilers, or co-firing the RDF with pulverized coal (Chapter IV). All such systems, however, are based upon various approaches to combustion fundamentals.

REFERENCES

1. Babcock and Wilcox. 1977. Steam: Its Generation and Use, 39th Ed. Babcock and Wilcox, New York.

2. Barton, R. G., Clar, W. D., Lanier, W. S., and Seeker, W. R. 1988. Dioxin Emissions During Waste Incineration. Presented at the Spring Meeting, Western States Section, The Combustion Institute.

3. Bradbury, A. G. W., and Shafizadeh, F. 1980. The Role of Oxygen Chemisorption in Low-Temperature Ignition of Cellulose. *Combustion and Flame*. *37*:85-89. Also see Bradbury, A. G. W, and Shafizadeh, F. 1980. Chemisorption of Oxygen on Cellulose Char. *Carbon*. *18*:109-116.

4. Bumb, P. R. et al. 1980. Trace Chemistries of Fire: A Source of Chlorinated Dioxins. *Science*. *210*(4468): 210-215.

5. California Air Resources Board. 1984. Air Pollution Control at Resource Recovery Facilities. Final Report. Sacramento, California.

6. Cowling, E. B., and Kirk, T. K. 1976. Properties of Cellulose and Lignocellulosic Materials as a Substrate for Enzymatic Conversion Processes. Enzymatic Conversion of Cellulosic Materials: Technology and Applications (E. G. Gaden et al., eds.), pp. 95-124. Wiley Interscience, New York.

7. Ebasco. 1982. Technology Assessment: Municipal Solid Waste as a Utility Fuel. Electric Power Research Institute, Palo Alto, California. (See particularly, R. E. Kaiser. 1975. Physical-Chemical Character of Municipal Refuse as referenced in Ebasco, 1982).

8. Edwards, J. 1974. Combustion: Formation and Emission of Trace Species. Ann Arbor Sci., Ann Arbor, Michigan.

9. Glassman, I. 1986. Combustion, 2nd Ed. Academic
    Press, San Diego, California.

10. Gordon, S., and McBride, B. 1971. Computer program for
    calculation of complex chemical equilibrium concentra-
    tions. National Aeronautics and Space Administration,
    Washington, D.C.

11. Hayes, B. W. et al. 1978. Sources of Metals in the
    Combustible Fraction of Municipal Solid Waste. U.S.
    Department of the Interior, Bureau of Mines, Washington,
    D.C. RI 8293.

12. Hazen Research. 1985. Personal communication from
    R. C. Hodgson to A. J. Rossi.

13. Hougen, O. F., Watson, K. A., and Ragatz, R. A. 1954.
    Chemical Process Principles, Part 1: Material and
    Energy Balances (2nd Ed.). John Wiley & Sons, New York.

14. Junge, D. C. 1975. Boilers Fired with Wood and Bark
    Residues. Forestry Research Laboratory, Oregon State
    University.

15. Junge, D. C. 1978. Design Guideline Handbook for
    Industrial Spreader Stoker Boilers Fired with Wood and
    Bark Residue Fuels. Oregon State University Press,
    Corvallis, Oregon.

16. Kaiser, R. E. 1975. Physical–Chemical Character of
    Municipal Refuse. *In* Proceedings of the 1975 Inter-
    national Symposium on Energy Recovery from Refuse.
    University of Louisville, Louisville, Kentucky.

17. Mulcahy, M. F. R., and Young, C. C. 1975. The Reaction
    of Hydroxy Radicals with Carbon at 298°K. *Carbon 132*(2):
    115–124.

18. Munro, J. M. 1982. Formation and Control of Pollutant
    Emissions in Spreader–Stoker–Fired Furnace. PhD Disser-
    tation, University of Utah, Salt Lake City, Utah.

19. Palmer, H. B. 1974. Equilibrium and Chemical Kinetics
    in Flames. *In* Combustion Technology: Some Modern
    Developments (H. B. Palmer and J. M. Beer, eds).
    Academic Press, New York.

20. Pershing, D. W., and Wendt, J. 1976. Pulvarized Coal
    Combustion: The Influence of Flame Temperature and Coal
    Composition on Thermal and Fuel NO_x. Proc. Sixteenth
    International Symposium on Combustion. The Combustion
    Institute.

21. Pershing, D. W. et al. 1978. The Influence of Fuel
    Composition and Flame Temperature on the Formation of
    Thermal and Fuel NO$_x$ in Residual Oil Flames. Proc.
    Seventeenth International Symposium on Combustion.
    The Combustion Institute.

22. Seeker, W. R., Lanier, W. S., and Heap, M. P. 1987.
    Combustion Control of MSW Incinerators to Minimize
    Emissions of Trace Organics. EER Corporation, Irvine,
    California for USEPA under Contract No. 68-02-4247.

23. Shafizadeh, F., and DeGroot, W. 1976. Combustion
    Characteristics of Cellulosic Fuels. *In* Thermal Uses
    and Properties of Carbohydrates and Lignins (F. Shaf-
    izadeh, K. V. Sarkanen, and D. A. Tillman, eds.).
    Academic Press, New York.

24. Shafizadeh, F., and DeGroot, W. 1977. Thermal Analysis
    of Forest Fuels. *In* Fuels and Energy From Renewable
    Resources (D. A. Tillman, K. V. Sarkane, and L. L.
    Anderson, eds.). Academic Press, New York.

25. Shaub, W. M., and Tsang, W. 1983. Dioxin Formation in
    Incinerators. *Environmental Science and Technology,*
    17(12):721-730.

26. Swedish Environmental Protection Board. 1986. Waste to
    Energy Report.

27. Tillman, D. A. 1980. Fuels From Waste. *In* Kirk-Othmer
    Encyclopedia of Chemical Technology, Vol. 11. Third Ed.
    pp. 393-410. John Wiley & Sons, New York.

28. Tillman, D. A., Rossi, A. J., and Kitto, W. D. 1981
    Wood Combustion: Principles, Processes, and Economics.
    Academic Press, New York.

29. Tillman, D. A., and Smith, W. R. 1981. The Evolution of
    Nitrogen Volatiles from the Pyrolysis of Red Alder Bark.
    Presented at the Forest Products Research Society Annual
    Meeting, St. Paul, Minnesota

30. Tillman, D. A., and Anderson, L. L. 1983. Computer
    Modelling of Wood Combustion with Emphasis on Adiabatic
    Flame Temperature. Journal of Applied Polymer Science:
    Applied Polymer Symposium 37:761-774.

31. Tillman, D. A. 1987. Biomass Combustion. *In* Biomass:
    Regenerable Energy (D. O. Hall, and R. P. Overend, eds.).
    pp. 203-219. John Wiley & Sons, Ltd., London.

Chapter III

MASS BURN SYSTEMS FOR COMBUSTION OF

MUNICIPAL SOLID WASTE

I. INTRODUCTION

Mass burning of mixed municipal solid waste (MSW) is the
dominant technology used for thermal volume reduction to
thermally destroy household refuse. Three basic components
are associated with the combustion process in MSW incineration systems: the waste receiving area (tip hall and pit)
and waste feed system, the combustion system itself, and the
air quality control system. The ash disposal system may be
considered as a second process associated with the mass burn
plant, and it may or may not include ferrous recovery from the
bottom ash, mixing of bottom and collected flyash (including
spent scrubber sorbent), and ash conditioning. A heat recovery and utilization system may be included as a third process
distinct from the incineration system. This system, if
implemented, consists of a make-up water system, feedwater
treatment and heating, the boiler system (including economizer,
boiler, and superheater), the turbine-generator, and the con-
denser and waste heat rejection system. Of these systems, the
combustion process of mass burning is the focal point of this
chapter.

This chapter is designed to provide a process overview of
mass burn systems focusing upon the most critical combustion
scenario. It includes an overview of the technology. Subsequently, case studies of mass burning at select locations
are considered and fundamentals associated with the process
are examined. The chapter concludes with a review of the
capital and operating costs associated with mass burning of
MSW.

A. Mass Burning Technology Overview

Practiced in many cities throughout the world, mass burning of MSW is considered a mature technology. Currently, the
United States has 74 waste-to-energy plants (see Table I)
with a combined operating capacity of 37,800 tons per day

TABLE I. Representative Existing Mass Burn
Facilities in the United States

Plant Location	Size (Tpd)	Supplier
Tuscaloosa, AL	300	Consumat
Hillsborough County, FL	1,200	Ogden–Martin
Panama City, FL	510	Westinghouse
Pinellas County, FL	3,000	Signal[a]
Tampa, FL	1,000	Waste Management
Baltimore, MD	2,250	Signal[a]
North Andover, MA	1,500	Signal[a]
Saugus, MA	1,500	Signal[a]
Claremont, NH	200	Signal[a]
Portsmouth, NH	200	Consumat
Peekskill, NY	2,250	Signal[a]
Tulsa, OK	1,125	Ogden–Martin
Marion County, OR	550	Ogden–Martin
Gallatin, TN	200	Westinghouse–O'Connor
Nashville, TN	1,120	Detroit Stoker/ Babcock and Wilcox
Rutland, VT	240	Vicon/Ingersoll Rand

[a] Signal Environmental Systems; now in joint venture with
Waste Management.

Source: Berenyi and Gould [3].

TPD [27]. This amounts to about 120 million tons per
year. The U. S. has almost 156,000 TPD of MSW-to-energy
capacity in operation, under construction, and in permitting.
Nearly 60% of this total capacity will be supplied by five
vendors: Ogden–Martin, Waste Management, Inc. (including the
joint venture with Signal, formerly Wheelabrator Frye or
Signal Environmental Systems), American Ref-Fuel, Combustion
Engineering, and Westinghouse Electric. Ogden–Martin and
Waste Management/Signal Environmental Systems, taken together,
command 37.3% of the market [27]. All of these developers

offer mass burn systems, and only Combustion Engineering
offers a non-mass burn (RDF) system.

In reality, mass burning is a family of technologies including basic types of systems: (1) the conventional grate-
fired waterwall incinerator, (2) the rotary kiln incinerator
using either refractory or waterwall kiln design, and (3) the
modular incinerator. All of these technologies are designed
to manipulate the combustion mechanism for MSW shown
previously (see Chapter II).

The largest number of mass burn installations are modular
incinerators, which may range in size from less than 10 TPD to
about 200 TPD. They may be installed by communities as well
as by large commercial waste generators. The units depend
upon batch feeding and semi-continuous feeding of the MSW
into the system. Heating and drying, pyrolysis, and char
oxidation occur in the primary chamber, while gas phase oxidation reactions occur in the secondary (or upper) chamber.
Modular units may be installed with or without heat recovery
systems. Typically, heat recovery systems are unfired waste
heat boilers which recover energy from the gaseous products
of combustion. The energy is in the form of low pressure,
process steam.

The preponderance of capacity for MSW incineration is
now, and is projected to be, concentrated in the large scale
waterwall and rotary kiln incinerators. These systems have
the capability to serve communities generating from 250 to
3,000 TPD of waste. Large scale systems can operate essentially on a continuous basis. In the waterwall incinerator
combustion systems (Fig. 1), waste is fed onto the grates by
means of a ram feeder. The waste undergoes heating and drying on the upper portion of the grate. Gas phase oxidation
reactions occur above the grate. Char oxidation reactions
occur at the "toe" of the grate just prior to the bottom ash
removal system. In rotary kiln systems, a heating and drying, and pyrolysis and ignition grate may be provided. The
rotary kiln may be used only to accomplish char oxidation
reactions. Alternatively, the raw refuse may be fed directly
into a kiln, and all reactions may occur with the rotating
reactor.

B. Process Analysis of Mass Burn Systems

This chapter focuses on the larger scale mass burn systems which dominate the current and projected MSW incineration capacity. The chapter examines the types of large scale
systems in terms of how they manipulate the combustion

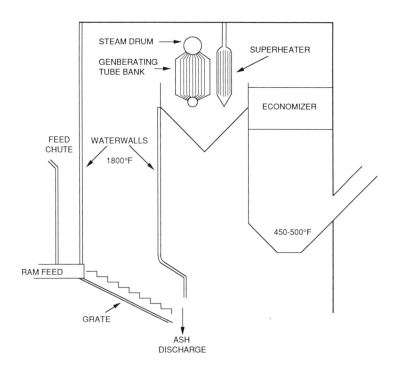

FIGURE 1. Schematic of a mass burn waterwall incinerator, including representative temperatures.

mechanism, what combustion parameters are used, what thermal efficiencies and flame temperatures are achieved, and what "uncontrolled" airborne emissions and solid residues are generated. The final section of the chapter considers the capital and operating costs associated with the mass burning of MSW.

Only those issues directly related to the process of MSW combustion are discussed in this chapter, including: heat release, thermal efficiency, and environmental consequence. Issues of energy utilization for electricity generation, process steam raising, or cogeneration are reviewed as applicable to small power plants. However, these latter issues are discussed only briefly as they are largely independent of the fuel being burned. The chapter also addresses the generation of certain airborne and solid emissions and residues but it does not consider the regulatory environment associated with those environmental concerns. Regulatory considerations vary substantially by location, and are consequently more appropriately addressed in site-specific analyses. To provide a useful assessment of mass burn technologies, representative plant descriptions and case studies are used throughout the remainder of the chapter. The case studies then support the more fundamental analyses of the mass burn approach to MSW incineration.

II. CASE STUDIES IN THE MASS BURNING OF MUNICIPAL WASTE

Five primary case studies represent the technological options available within the family of mass burn technologies. These case studies include three mass burn waterwall incinerators and two rotary kilns. The waterwall incineration case studies reflect the variety of grate-types developed in Europe and the United States, and the changes that occur as a function of system size and date of installation. The rotary kiln case studies include a refractory-lined kiln and a waterwall kiln.

A. Case Studies of Mass Burn Waterwall Incinerators

The waterwall incineration case studies include facilities in Hogdalen, Sweden; Nashville, Tennessee; and Pinellas County, Florida. The Hogdalen plant is a most recent installation with the addition of Boiler No. 3 in 1986. The Nashville plant was one of the first three modern waterwall incinerators installed in the U.S. while the Pinellas County plant is the largest mass burn facility in the U.S. These case studies are expanded by discussions of the mass burn waterwall incinerators in other U.S. and European cities.

1. *The Hogdalen, Sweden, MSW Incinerator*

The Hogdalen, Sweden, plant is a cogeneration facility in
south Stockholm. It currently consumes 220,000 tons per year
(200,000 tonne per year) of MSW and produces both electricity
and hot water for district heating. It is a three boiler
installation. The first two boilers are 11 tons per hour
(10 tonne/hr) units with VKW–Deutche Babcock roller grates.
The third boiler is a 16.5 ton per hour (15 tonne/hr) unit
utilizing a Martin GMBH grate shown in Fig. 2. Boiler No. 3
was installed in 1986. The Hogdalen boiler No. 3 produces
steam at 540 psig/680°F (36 atm/360°C) for use in cogeneration of electricity and district heat. All three boilers are
served from a tipping hall capable of handling 12 trucks at a
time. The tipping hall opens to a single, oversized pit,
capable of handling waste deposited for a full week. Two

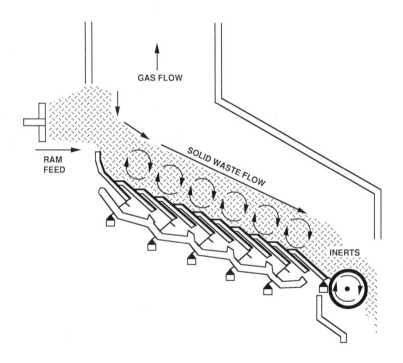

*FIGURE 2. Martin reverse reciprocating grate used
in Hogdalen Boiler No. 3.*

cranes mix the waste in the pit, and then feed it through chutes to the waste fired boilers. Boiler No. 3 reflects the latest developments in technology, and is the focus for the remaining discussion of the Hogdalen facility [24].

a. *Combustion at the Hogdalen Facility.* Boiler No. 3 is fed by a 20-foot (6-meter) high chute. The chute is filled with MSW to maintain a good seal, to prevent air from flowing into the boiler from the fuel feed system, and to prevent fire from burning into the feed system. Fuel is fed to a Martin reverse reciprocating grate both simultaneously and sequentially via a ram feeder. The active grate surface is about 33 feet long (10 meters) and accomplishes the following general phases of solid fuel combustion: heating and drying, solid particle pyrolysis, ignition, and solid carbon burnout (Fig. 2). The mass of fuel enters at the top of the grate where it is dried and pyrolysis begins. Ignition and combustion occur on the midpoint of the travel path on the grate. Burnout occurs as the refuse moves towards the toe of the grate. The reverse reciprocating action of the grate causes the burning fuel to be pushed against the flow of gravity so that burning fuel is pushed under the mass of incoming fuel. This action facilitates complete MSW burnout. Further, air is distributed independently to various portions of the Hogdalen Boiler No. 3 grate as shown in Fig. 3 to ensure correct combustion stoichiometry and complete oxidation of the MSW. At Hogdalen, solids remain on the grate for 1.5 hours before being deposited in the wet ash handling system [24].

The critical parameters of the Hogdalen Boiler No. 3, summarized in Table II, include furnace height, furnace design, and the consequent combustion process. The furnace height of the unit is approximately 65 feet (20 meters) and there are three passes in the unit. The first 32 feet (10 meters) of the furnace is refractory lined. Gases passing through the furnace have a residence time at temperatures exceeding 1650°F (900°C) for at least 2 seconds (Fig. 4). Total gaseous residence time in the unit is up to 5 seconds. Gaseous velocity in the unit was specified at less than 16 ft/sec (less than 5 m/sec) which compares to 35 ft/sec (11 m/sec) for Hogdalen Boilers No. 1 and No. 2. The Hogdalen Boiler No. 3 operates at a calculated thermal efficiency of about 72% (Fig. 5). It has a carbon conversion efficiency of 98 to 99%, compared to the carbon conversion efficiency of 95% for the other boilers [24].

b. *Airborne Emissions at the Hogdalen Facility.* The Hogdalen Boiler No. 3 generates a moderate amount of particulates since 78% of the unburned solids and inerts are passed

*FIGURE 3. Air distribution system installed at the
Hogdalen, Sweden, Boiler No. 3. Note the four distinct
locations for combustion air. Note also that the second
and third ducts have the largest air carrying capacity.*

from the unit as bottom ash and only 22% are discharged as
flyash. Uncontrolled particulate emissions are approximately
3.15 gr/dscf (at 12% CO_2). Dioxin control in Hogdalen
Boiler No. 3 is achieve by controlling the level of excess
air and air distribution. These lead to a grate stoichiometric ratio of 1.19, and a consequent avoidance of reducing
conditions within the unit. Dioxin control is also achieved
by combustion temperatures exceeding 1800°F (982°C) for
greater than 2 seconds (Fig. 4), and by the use of an advanced
air quality control system which includes dry acid gas sorbent
injection and a baghouse, and low AQCS temperatures
(i.e., less than 340°F) for airborne emission minimization
(see Chapter VIII). The Hogdalen plant, then, uses a combination of good combustion control and an advanced post-
combustion AQCS to minimize airborne emissions [24].

TABLE II.   Some Critical Combustion Parameters at
      Hogdalen, Sweden

Parameter	Value
Feed rate	16.5 ton/h (15t/h)
	152×10^6 Btu/h (40.5 MW_t)
Grate	Martin
Grate dimensions	32 1/2 ft(1) × 16.3 ft (10m × 5m)
Heat release	297×10^3 Btu/ft^2 h (.92 MW/m^2 grate)
	8.3×10^3 Btu/ft^3 h (.083 MW/m^3 furnace)
Residence time, solids	1 1/2 hr
Residence time, Gas σ T>1,800°F	2 sec
Excess air	
As excess Air	78.5%
As dry O_2 in stack	9.3%
Air distribution	
Primary	70.0%
Secondary	30.0%
Combustion air temp.	77°F (25°C)
Adiabatic flame temp.	2,180°F (1,190°C)
Final useful stack temp. (Not including condensing econ. in AQCS)	365°F (185°C)
Carbon conversion	98–99%
Percent ash as flyash	22%
Percent ash as bottom ash	78%

Source: [24].

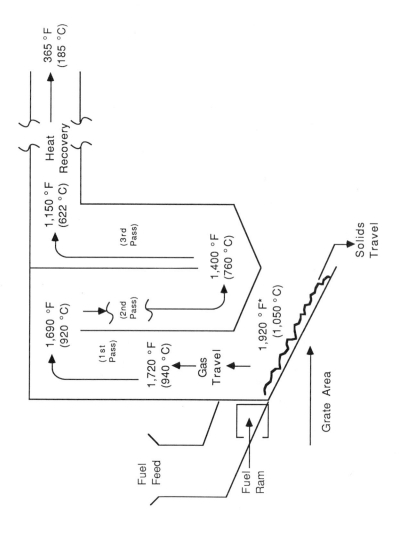

* Adiabatic Flame Temperature is 2,180° F (1,190° C)

FIGURE 4. Temperature profile of Hogdalen No. 3 boiler.

68

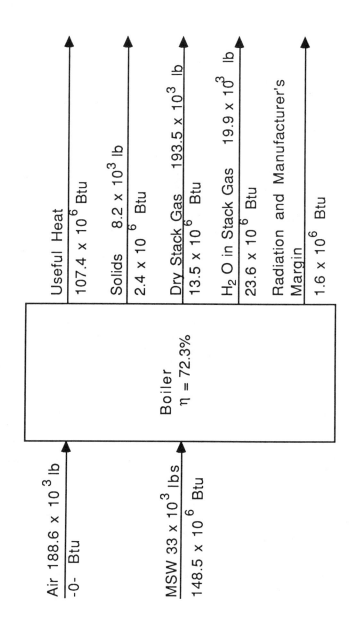

FIGURE 5. Calculated heat balance about Hogdalen, Sweden, No. 3 Boiler.

69

2. The Nashville, Tennessee, MSW Incinerator

The Nashville Thermal Transfer Facility was originally
installed in 1974 as a 720 TPD facility. It was expanded to
three combustors and a capacity of 1,120 TPD in 1984. Refuse
is brought to the waste processing building (Fig. 6) and
deposited in the pit. Large cranes then lift the refuse and
place it in one of three charging hoppers. On demand, the
refuse is forced onto the Detroit stoker grate system by a
ram feeder (see Fig. 7). The inclined grate system at
Nashville Thermal Transfer has distinct sections for drying,
ignition and burning, and final burnout as shown in Fig. 7.
Combustion occurs both on and above the grate. Products of
combustion pass through radiant, convection, and economizer
sections of the boiler before being treated in an electrostatic precipitator. Following particulate removal, the
gaseous products of combustion are discharged to the atmosphere. Bottom ash and flyash are collected and disposed of
in a landfill. This incineration facility supplies direct

FIGURE 6. Waste processing building at Nashville
Thermal Transfer Plant. Note electrostatic precipitators and stacks at far left.

FIGURE 7. Ram feeder and grate profile at Nashville
Thermal Transfer. Note the distinct steps on the grate
to promote waste tumbling and complete combustion. Also
note overfire air jets to facilitate gas phase oxidation.

heating steam and chilled water to buildings in downtown
Nashville and generates some 7.5 megawatts (MW) of electricity
for sale to the grid [1, 9, 11, 24].

        a. Combustion at the Nashville Thermal Transfer Facility.
The Detroit stoker grates in Nashville Thermal Transfer units
No. 1 and No. 2 are 12 feet wide and 53 feet 4 inches long
with surface areas of 630 square feet. The Detroit stoker
grate in Unit No. 4 is 15 feet wide and 39 feet long [9].
Unit No. 3 is a standby fossil fuel boiler.) The grates in
units No. 1 and No. 2 are designed for a feed rate of about
50 lb/sq ft/hr, and a heat release rate of about 220,000
Btu/sq ft/hr. The feed rate for unit No. 4 is 65 lb sq ft/hr,
and the heat release rate is 292,500 Btu/sq ft/hr. The furnaces have a volumetric heat release rate of about 10,000
Btu cu ft/hr [9].
        Made of a high temperature nickel alloy stainless steel,
the grates are set at an angle of about 15 degrees. Distinct elevation breaks in the grates introduce a tumbling

action and increase the completeness of combustion. Further, the breaks provide some distinction between phases of the combustion process and physical areas where manipulation of the combustion mechanisms takes place. Heating and drying occurs on a charging grate while pyrolysis/gasification, flaming combustion, and burnout takes place on and above two combustion grates. A solids burnout grate provides for char oxidation reactions [24].

Combustion conditions associated with the Detroit stoker system include the use of 80% excess air and a distribution of 60% underfire/40% overfire air [9,11]. Combustion air is preheated only when required for fuel drying; otherwise, ambient air is used. Consequently, the Nashville plant has an approximate average grate stoichiometric ratio (GSR) of 1.063, and a calculated adiabatic flame temperature of 2160°F (1180°C). The approximate temperature profile of the Nashville system includes a furnace temperation of 1800 to 2100°F (980 to 1150°C), a furnace exit temperature of 1500°F (820°C), a convection section exit temperature of 760°F (400°C) and an economizer exit temperature of 450°F (232°C) [24]. These combustion conditions yield an approximate thermal efficiency of about 68%.

   b. *Airborn Emissions at the Nashville Thermal Transfer Facility.* The Nashville Thermal Transfer Facility has three MSW boilers coupled to a four-field electrostatic precipitator (ESP). Each precipitator is capable of 99% particulate collection efficiency. No post-combustion control system has been installed for sulfur dioxide (SO_2), nitrogen oxides (NO_x), or other criteria pollutants [1, 15, 24].

Resource Consultants performed stack testing on unit No. 2 in February, 1983. The results of those stack tests included particulate emissions of 0.022 gr/DSCF corrected to 12% CO_2 after the ESP. SO_2 emissions were 77 ppmv, NO_x emissions were 150 ppmv, and CO emissions were 70 ppmv [15]. Organic emissions from the facility, including dioxins and furans, have not been determined. However, the temperature and air distribution profiles discussed previously indicate a regime capable of minimizing their presence. The incineration system at Nashville accomplishes a 78% reduction in the mass of solid waste, and a 90% volumetric reduction in the MSW stream [25]. With the system now fourteen years old and an availability in excess of 85%, the facility demonstrates the utility of mass burn incineration as a waste disposal technique.

3. The Pinellas County, Florida, MSW Incinerator

The Pinellas County plant, shown in Fig. 8, is among the largest mass burn facilities constructed to date, having a waste incineration capacity of 3,000 TPD and an electricity generating capacity of 75 MW. Refuse is delivered by truck and dumped in a receiving building. Two large overhead cranes deposit the waste in feed chutes for the three large refuse boilers. Each boiler has the capacity to burn 1,000 TPD of refuse and generate some 702,000 lb/hr of 600 psig/ 750°F (41 atm/400°C) steam. The steam is used for generating electricity in one 50 MW turbine and in a second 25MW turbine. Gaseous products of combustion are passed through three-field ESPs, where particulates are moved. Bottom ash is transported to a ferrous metals recovery section. All ash is landfilled. The Pinellas County facility typically processes 2,500 TPD of refuse and achieves a 90–95% volumetric reduction of the waste [21, 24].

FIGURE 8. Pinellas County, Florida, 3,000 TPD MSW-to energy facility. Note the use of outdoor installation of boilers, turbine-generators, and ESPs.

   a. Combustion at the Pinellas County, Florida, Facility.
Combustion occurs on a Martin reverse reciprocating grate,
which is the most popular grate system in current mass burn
facilities. A ram feeder forces refuse onto the grate where
it undergoes heating and drying, ignition, combustion, and
burnout. The grate is short, smooth, and set at an angle of
26 degrees. Consequently, all combustion reactions occur
somewhat simultaneously. The mixing of the solids is achieved
by the reverse reciprocating action of the grate bars. Gravity
pulls the refuse down the grate; however, the grate bars push
the burning refuse against the flow of incoming material.
Burning refuse is pushed under colder material to promote its
heating, drying, and ignition (Fig. 9). The grate length at
Pinellas County is 17.7 feet, which was standard for that
generation of Martin grates (see Turner [25]). Furnace
capacity is increased by adding to the width of the unit as
shown in Fig. 10 [25].

*FIGURE 9. View of the Martin grate at Pinellas County,
Florida. Note the ram feeder in the upper portion of
the picture and the ash discharge mechanism in the foreground. Also note the absence of breaks or drops as
designed into the Detroit stoker grate and the severity
of grate angle.*

FIGURE 10. Cross sectional view of Pinellas County MSW boiler. Note the ram feeder, the short grate length, the severity of grate angle, and the positioning of grate bars. Also note the use of refractory in the furnace section to protect the waterwalls.

The grate system at Pinellas County is approximately 800 square feet per boiler and is used to combust primarily residential refuse with about 4,200 Btu/lb (2,333 cal/g), and 30% moisture. Solids residence time on the grates is approximately 20 minutes. Consequently, the grate heat release rates are about 410,000 Btu/sq ft/hr [24]. The furnace is about 50 feet high with a volume of about 36,000 cubic feet. The furnace length and width are not equal to the grate dimensions due to the slope of the grate. The volumetric heat release rate is about 9,100 Btu/cu ft/hr [24].

The Pinellas County system operates with an air distribution of 70% undergrate/30% overfire air [24]. The excess air level is 90% (calculated from 10% O_2 in the stack gas), yielding an equivalence ratio of 0.526 and a grate stoichiometric ratio of 1.259. Combustion air is moderately preheated to 200 to 250°F (93 to 121°C)[24].

The approximate temperature profile of the Pinellas County unit demonstrates that it achieves temperatures

greater than 1800°F (982°C) for 1 to 2 seconds as shown in
Fig. 11.   Combustion at the facility is considered complete,
with the carbon content in the ash ranging from 0.2 to 0.8%.
The thermal efficiency of the boilers exceeds 70% [24].

        b.   Airborne Emissions at the Pinellas County Facility.
Airborne emissions are uncontrolled except for particulates.
Data on dioxins and furans are not available; however, the

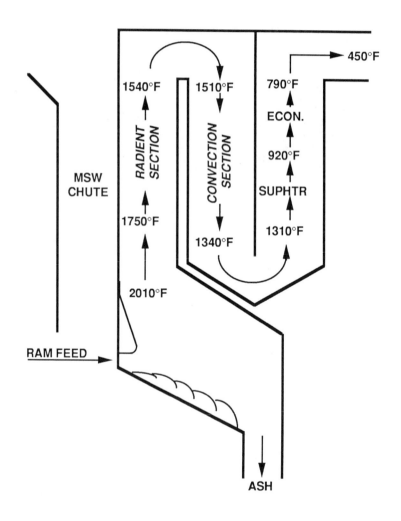

Figure 11.   Temperature profile of the Pinellas County
Plant.   Source: [24].

combustion regime would indicate good control. Carbon monoxide emissions can exceed 200 ppmv (dry stack gas) as can
oxides of nitrogen emissions [24]. Sulfur dioxide emissions
are 61 ppmv (dry stack gas) [24]. The Pinellas County facility has an operating availability of 90% and a typical capacity factor of 85% when measured on a tons-processed basis, or
87% when measured on an electricity-generated basis [21, 24].

4. Summary Data From Other Operating MSW Incinerators.

This section provides a comparison of the plants discussed above to certain waterall facilities, including the
1,500 TPD facility in North Andover, Massachusetts; the
2,250 TPD facility in Baltimore, Maryland; and, the 850 TPD
facility in Uppsala, Sweden. The North Andover facility is a
two-boiler unit with Martin grates. Each furnace is identical
in size to the Pinellas County furnaces [19]. However,
because of very rich fuel, the North Andover plant has a significantly lower tonnage throughput per boiler than the
Pinellas County plant. MSW processed at the plant contains a
heat value of 5,100 Btu/lb (2,833 cal/g) compared to its design waste stream of 4,600 Btu/lb (2,555 cal/g) and the Pinellas County waste stream of 4,200 Btu/lb (2,333 cal/g). Operating parameters are also slightly different than those at
Pinellas County. North Andover operates with slightly less
excess air (8.5 to 9% O_2 in the stack); and with somewhat
hotter flame temperatures that typically achieve 2300°F
(1260°C). Its airborne emissions are as follows: CO at 20
to 25 ppmv (corrected 12% CO_2); NO_x at 0.4 lb/Btu × 10^6; and
SO_2 at 0.15 to 0.2 lb/Btu × 10^6. Carbon conversion is in the
98 to 99% region. Plant availability for this 40 MW facility
is 91% [24].

The Baltimore, Maryland, facility is a 60 MW power plant
comprised of a three-boiler unit. Each furnace is 25.5 feet
wide by 33.6 feet long and capable of burning 750 TPD. The
plant uses Von Roll Reciprocating grates which are set at an
18 degree angle. The grate heat release is about 330,000
Btu/sq ft/hr and the volumetric heat release rate is about
10,000 Btu cu ft/hr. The unit operates at about 85% excess
air. The Baltimore plant is a sister facility to the Westchester, New York, facility. However, the Westchester facility has a somewhat higher firebox temperature than the other
mass burn facilities, demonstrating flame temperatures in
the 2100 to 2500°F (1150 to 1370°C) range [19, 20]. Some air
preheating may be used for underfire air. Gas burners in the
furnace maintain high temperatures if unusually wet or low
Btu content waste is processed. Temperatures are held in
excess of 1800°F (982°C) for at least 2 seconds [24].

The Uppsala, Sweden, plant is a four-boiler system using Wiedmar and Ernst horizontal reciprocating grates. The system is operated 8,200 hr/yr. Two boilers burn 3.9 ton/hr (3.5 tonne/hr), one boiler burns 11 ton/hr (10 tonne/hr) and the other boiler burns 16.5 ton/hr (15 tonne/hr). The grate bars are air cooled, and set at a 0° slope. Waste moves across the unit as a result of grate bar action. The largest unit has a grate which is 49 feet (15 meters) long by 13 feet (4 meters) wide, and has a consequent slow grate heat release rate of 233×10^3 Btu/sq ft per hour. It has relatively low combustion gas velocities, and combustion temperatures of about 2010°F (1100°C) according to the plant operators. Carbon conversion is on the order of 95%. Excess air levels are low, with excess O_2 in the stack being about 6.5%. The air distribution is 60% primary air and 40% secondary air [24].

These case studies demonstrate that mass burn waterwall incineration is a well developed, mature technology with substantial potential as a waste management strategy. The technology is capable of accomplishing a 70 to 80% reduction in the mass of waste going to landfill, and a 90 to 95% reduction in the volume of waste going to landfill. At the same time mass burn waterwall incinerators can generate substantial quantities of useful energy, largely as electricity, while producing minimal amounts of airborne emissions.

B. Case Studies in Rotary Kiln Mass Burn Incinerators

Alternatives to mass burn waterwall incineration technologies are the refractory-lined and waterwall rotary kilns, developed by Volund and Westinghouse-O'Connor, respectively. The Waste Management, Inc., facility at Tampa, Florida, uses the Volund system, and the Sumner County, Tennessee, facility uses the Westinghouse-O'Connor combustors.

1. *The Tampa, Florida, MSW Incinerator*

The Tampa, Florida, plant is also known as McKay Bay. It is a 1,000 TPD system that uses a Volund design rotary kiln to generate 25 MW of electricity, a design which had previously been installed in Copenhagen, Denmark. Four process lines at the plant each provide a maximum 250 TPD of capacity. Actual process rates for 1986 averaged 225 TPD, based on 277,711 tons of refuse processed [24, 26].

An overview of the McKay Bay facility is shown in Fig. 12. The refuse loading cranes which feed the four waste processing lines are operated from the boiler control room as

FIGURE 12. Overview of McKay Bay facility. Note the
relationship between the waste processing building (far
right) which includes the pit, control room, and ignition grates; the kilns; and the waste heat boiler
(left).

shown in Fig. 13. The crane pulpit is a projection from the
control room, which facilitates communication between all
operators.

There are several components to the combustion system
including dyring and ignition grates, the rotary kilns, and
furnace sections. The drying and ignition grate sections of
the furnace are depicted in Fig. 14. The heart of the system
are the rotary kilns shown in Figs. 15 and 16. Gaseous
products of combustion pass through the waste heat boiler or
heat recovery steam generator (HRSG), which generates 650
psig/700°F (45 atm/370°C) steam. The McKay Bay facility
typically generates 22.5 MW of electricity [24, 26].

a. Combustion at the Tampa, Florida, Facility. The fuel
burned at the plant is a mixed MSW with about 30% moisture,
20 to 24% ash, and an average of 4,500 Btu/lb (2,500 cal/g).
This MSW has a heat content range of 3,000 to 6,000 Btu/lb

FIGURE 13. McKay Bay waste pit as viewed from the pulput extending from the control room.

FIGURE 14. Drying and ignition grate section of Volund Incinerator Technology installed in Tampa, Florida.

(1,667 to 3,334 cal/g) depending upon seasonal factors and moisture. A ram charger feeds the refuse onto a drying grate where it is heated and dried, and pyrolysis and gasification reactions are initiated. The refuse then tumbles to the ignition grate where combustion begins. Each process line has a heat release of 94 × 10⁶ Btu/hr and an effective grate surface of 382 square feet. This yields a grate heat release rate of about 250,000 Btu/sq ft/hr. Each kiln is 12 feet 2 inches in diameter, and 23 feet long, with a consequent length/diameter (L/D) ratio of 1.89. Low L/D ratios minimize particulate entrainment and flyash generation. Each kiln has about 880 cubic feet of volume, or 13% of the total process line furnace volume (6,700 cubic feet). The Volund units at McKay Bay have furnace heat release rates of 14,000 Btu cu ft/hr and a solids residence time of 2 hours and 40 minutes. This solids residence time can be varied by altering the rotational speed of the kiln from 0 to 12 revolutions per hour [24].

FIGURE 15. Volund rotary kilns installed at the McKay Bay facility. These kilns discharge to the heat recovery steam generator.

*FIGURE 16.   Close up of a Volund kiln at the McKay Bay
facility.*

The combustion process itself is manipulated by using the
discrete sections in the furnace (Fig. 17).   Solids are heated,
dried, and pyrolyzed on the grate system.   Volatiles ignite
and are combusted above the grate.   The gaseous products of
combustion pass directly to the HRSG.   Char oxidation occurs
in the refractory-lined rotary kiln, with the heat evolved
being used both in support of the combustion process and to
generate steam in the HRSG.   The kiln provides for solids
agitation (turbulence) and complete carbon burnout.   The McKay
Bay facility uses about 90% excess air (10% O_2 in the stack
gas), and has an air distribution of 60% undergrate/40% overfire air.   There is no direct air supply to the kilns,
although the system uses air preheat when fuel conditions
warrant it.   Frequently, the facility uses ambient air.   The
system has a GSR of about 1.10 and maintains gaseous temperatures in excess of 1800°F (982°C) for over 2 seconds as shown
in Fig. 17.   Gases exit the HRSG at 400°F (204°C) before being
ducted to the ESP.   These conditions provide for a thermal
efficiency of about 70% [24].

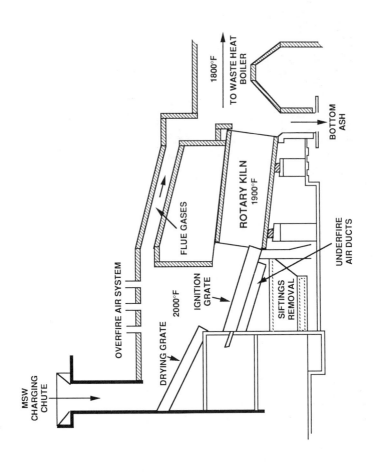

FIGURE 17. Schematic and temperature profile of the McKay Bay facility. Source: [24].

   b. *Airborne Emissions and Solid Residues from the Tampa,*
Florida, Facility. The facility is designed to minimize the
generation of uncontrolled airborne emissions, as indicated by
the long solids residence time and the very low L/D ratio of
the kiln.

   The unit achieves a 75% mass reduction and a 90-95% volume
reduction in the solid waste stream. The solids are virtually
free from combustibles, containing only 1 to 2% carbon. Of
the total solids discharged from the system, 17.6% are collected flyash and 82.4% are bottom ash (mass basis). Consequently, on an hourly basis each process line generates about
3,500 lb of bottom ash and 750 lb of collected flyash. The
stack gas, once cleaned by the ESP, contains a maximum of
0.025 gr/dscf at 12% CO_2. Bottom ash is collected and
processed for ferrous metals removal prior to disposal in a
landfill [24].

   Gaseous airborne emissions include CO at 30 ppmv, NMHCs at
1 ppmv (wet basis), NO_x at 70 ppmv (dry basis) and SO_2 at
80 ppmv. Although data are not available concerning dioxin and
furan emissions, the combustion regime described previously
offers evidence that such emissions would be controlled. The
required temperatures, residence times, and grate stoichio-
metries discussed previously in Chapter II are satisfied [24].
The McKay Bay facility has an availability exceeding 82%
(current operating records indicate an availability of about
90%). The capacity factor of the plant for 1986 was 76%
(277,711 tons processed/365,000 tons maximum potentially
processed) [24, 26].

2. *The Gallatin, Tennessee, MSW Incinerator*

   At 250 TPD, the Westinghouse-O'Connor rotary waterwall
kiln installed at Gallatin, Tennessee, is much smaller than
the systems previously discussed. The same technology, however, has been designed and installed in larger units, such
as the 510 TPD facility in Panama City, Florida. The Gallatin
plant has two processing lines, each capable of handling 125
TPD or 5.2 TPH of waste. Like all mass burn plants, grapple
cranes are used to load feed chutes connected to the furnaces
(in this case the kilns). Waste then enters the kiln shown
in Fig. 18, undergoes combustion, and the solids and gaseous
products are discharged from the lower end of the unit. The
gaseous products enter the HRSG, also shown in Fig. 18, where
steam is raised for electricity generation. Boiler tubes also
comprise the kiln walls for additional heat recovery. Gaseous
products of combustion are ducted to an ESP prior to their
discharge to the atmosphere [24].

FIGURE 18.   The Westinghouse-O'Connor kiln at Gallatin,
Tennessee. Note the discharge end of the kiln entering
the HRSG.

a. *Combustion at the Gallatin, Tennessee, Facility.*
Because the Westinghouse-O'Connor system has no grate, it is
clearly distinguished from other mass burn systems. The combustion mechanism is manipulated quite differently from the
systems described previously. Solids enter the kiln and
remain there for residence time of 35 minutes. The kiln is
set at an angle of 6 degrees and is operated at a rotational
speed of 3.7 RPH. Each kiln at Gallatin has a length of 28
feet and an inside diameter of 8.5 feet, yielding an L/D
ratio of 3.3. There is no grate heat release rate. Volumetric heat release is estimated at about 6,000 Btu/cu ft/hr
including the kiln and the final burnout section of the HRSG.
All stages of the combustion mechanism occur simultaneously
and sequentially in the kiln [24].

The rotary kilns at Gallatin operate on 50% excess air
(7.2% O_2 in the stack gas). Air is preheated to about 350°F
(230°C) and distributed within the kiln in three zones as
shown in Fig. 19. The general temperatures reported at

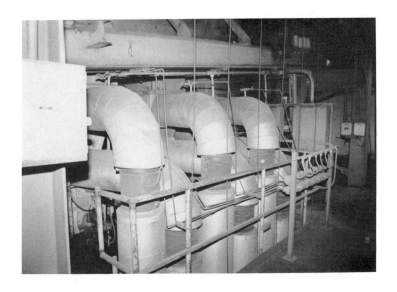

FIGURE 19. *Combustion air distribution system for the*
Westinghouse-O'Connor kilns in Gallatin, Tennessee.
Note the ducting of air to three separate zones in
the kiln.

Gallatin are as follows: fireball, 2100 to 2300°F (1150 to
1260°C); kiln exit, 1400°F (760°C); and boiler, 1150°F (620°C)
[24].   This temperature profile is somewhat complex, however,
due to the presence of cold waterwalls in the kiln.   Consequently, the fuel is periodically exposed to high temperatures
and then cooler wall temperatures.

     b.   Airborne Emissions and Solids Residues.   Airborne
emissions from the Gallatin facility were measured by Cooper
Engineers for the California Waste Management Board [6].
Results showed uncontrolled emissions as follows: particulates, 2.92 gr/dscf (corrected to 12% CO_2); NO_x, at 147 ppmv,
SO_2 at 154 to 182 ppmv, CO at 540 ppmv, and hydrocarbons at
40 ppmv (all corrected to 7% O_2.   Tests conducted for the
Electric Power Research Institute [13] demonstrated that the
bottom ash and collected flyash contained 11 to 11.6% carbon,
indicating a carbon conversion of only 86 to 87%.   Since the
tests of 1984, Westinghouse and Sumner County have significantly upgraded the Gallatin facility.   Their upgrade included
installing a new air distribution system for control of underfire and overfire air, a larger induced draft fan, complete
instrument recalibration, automated controls, new stacks,
water jackets on the feed chutes, and air preheat controls
[24].   Those improvements, along with upgrading of the operating practices, have reduced unburned carbon losses to less
than 5%.   Uncontrolled flyash and bottom ash also have been
reduced.   Combustion also has improved.   The unit now emits
80 ppmv of CO (83 ppmv when corrected to 7% O_2).   No new data
have been developed concerning other emissions [24].
     The waterwall rotary kiln, then, offers a significantly
different approach to mass burning of MSW.   It is an alternative that has gained market acceptance in Japan and is now
gaining market acceptance in the United States.

     c.   Summary of Case Studies.   The case studies discussed
above provide critical insights into the mass burning of
municipal waste.   There are a wide variety of designs, largely
focusing upon different approaches to the grate and combustion
chamber (Table III).   These designs include inclined and horizontal grates, reciprocating and reverse reciprocating grates,
drum roller grates, and kilns.   There are additional grate
designs that will be discussed subsequently.   However, among
these grates are reciprocating designs similar to the Detroit
Stoker and Von Roll systems discussed in the case studies.
Such alternative reciprocating grates would include the Riley-
Takuma (offered in the United States by Riley-Stoker Corporation), de Bartolemeis (licensed by Combustion Engineering),
and L. C. Steinmuller systems.   More distinct grate designs

TABLE III. Representative List of Mass Burn Grates
Available

Grate Type	Supplier
Reciprocating grate	Von Roll
	Detroit Stoker
	Riley Takuma
	L. C. Steinmuller
	de Bartolemeis
	Widmar and Ernst
Reverse reciprocating grate	Martin GmbH
Roller grate	Deutche Babcock
Rocking grate	
Circular grate	
Rotary kilns	Volund
	Westinghouse - O'Connor

Sources: [5, 17].

include circular grates, rocking grates, and rolling grates
[5]. Again, these are designed to agitate and mix the solids
being burned. Roller grates such as the VKW grates are noteworthy because the grate sections turn in the same direction
and consequently there is considerable agitation where the
rolls meet. One roll is directing the refuse downward, while
the second roll achieves a lifting action. All of these
grate systems are designed to achieve specific combustion
process objectives. Manipulation of combustion process variables for efficiency, environmental impact minimization, and
economic effectiveness is review below.

All technologies are capable of achieving mass reductions
of about 70%, and volumetric reductions of MSW on the order of
85 to 95%. Further, all technologies are capable of operating
at combustion temperatures of about 2000°F (1090°C), and
maintaining temperatures of greater than 1800°F (980°C) for
residence times of 1 to 2 seconds. All technologies achieve
thermal efficiencies of about 70%, with efficiencies largely
depending upon the temperature of the gaseous products of
combustion exiting the heat recovery system, rather than upon
the approach to the combustion process per se. Finally, all
mass burn designs are capable of minimizing airborne emissions
such as particulates, oxides of nitrogen, and organic

emissions (CO, hydrocarbons, and dioxins and furans) when
properly operated. Given these basic parameters, it is
useful to evaluate the fundamentals of the mass burn approach
to MSW incineration.

III. MANIPULATING THE PROCESS OF MASS BURN INCINERATION
      OF MUNICIPAL WASTE

     Control and manipulation of the process of combustion in
mass burn installations involves capitalizing on the three Ts
of combustion: temperature, time, and turbulence. These
three Ts of combustion are controlled in order to maximize
system efficiency, minimize environmental consequences of
municipal waste burning, and optimize capital utilization.
The following paragraphs evaluate the fundamentals of MSW
combustion, recognizing the wide variety of grate types, specific designs, and vendors serving the mass burn segment of
the MSW market.
     Two design issues generally dominate the combustion process
considerations of the specific alternative mass burn systems:
(1) the quantity, distribution, and temperature of combustion
air within the mass burn system; and (2) rates of heat release
on the grate and in the furnace of the mass burn system.
Combustion air manipulation largely impacts the completeness
of combustion, combustion and thermal efficiency, and environmental consequence issues. Rates of heat release also affect
these concerns through the residence time variable but have a
major impact on capital utilization as well. The issues of
combustion, air manipulation, and heat release are addressed
in succeeding sections of this chapter, integrating case study
data into more generalized information. The data available
on existing specific alternative designs such as those identified in Table III indicate that there is considered similarity
concerning these factors regardless of specific grate or
combustion system desitn.

A. Manipulation of Combustion Air

     The manipulation of combustion air dominates the combus-
tion parameter issue. Issues include percentage of excess
air, distribution of air, and temperature of combustion air.

1. Excess Air

     Excess air has a significant influence of flame tempera-
ture, thermal efficiency, CO emissions, and dioxins and
furans. Seeker et al. [17] have shown generally that excess

air levels of 40 to 135% (6 to 12% O_2 in the stack gas) are
required for MSW combustion; the data developed by Seeker [17]
convert to equivalence ratios of 0.714 to 0.425 as suitable for
mass burning of MSW. Operating data from well designed and
effectively operated plants generally confirm the broad range
of excess air levels posited by Seeker et al., and focus on a
much narrower segment of excess air levels in the middle of
that range. Typical excess air percentages cited are in the
80 to 110% range (equivalence ratios of 0.555 - 0.476) as
shown in Table IV.

It is useful to note the convergence of these excess air
levels, particularly when the Westinghouse excess air level
is disregarded. These data are further confirmed by recent
mass burn proposals provided by vendors to cities and coun-
ties interested in installing new mass burn facilities. Given
these ranges of acceptable excess air, it is useful to consider the specific combustion influences of this parameter.

   a.   *Influences of Excess Air on Flame Temperature and
Efficiency.* The influence of excess air on flame temperature
and efficiency can be calculated explicitly from equations
presented in Chapter II (see equations 2-30 through 2-35).
For this discussion, however, approximation equations have
been calculated based upon the generic ultimate analysis of
municipal waste previously shown, with 25.6% carbon, 3.4%
hydrogen, 20.3% oxygen, 24.4% ash, 25.2% moisture, and
4500 Btu/lb. (2.5 kcal/kg). These approximations assume the
conditions of the Hogdalen boiler, other than excess air.

TABLE IV.   Recommended Excess Air Levels for Mass
             Burn Systems

Mass burn grate system	Recommended excess air (%)	Ratio	Information (Literature) Source
Martin	90	0.526	Scherrer and Juran [16]
Von Roll	80	0.556	Plant visits [24]
Bartolemeis	85-110	0.541-0.476	Hazzard and Tease [8]
Detroit Stoker	80	0.556	Johnson and Reschly [9]
Volund	90	0.526	Tampa Plant Visit [24]
Westinghouse	50	0.667	Plant Visit [24]

Adiabatic flame temperature (°F) can be approximated
using the following equation:

$$T_{ad} = 86.567 + 3738(\phi) \tag{3-1}$$

Where ϕ is equivalence ratio. As shown in Chapter II, equivalence ratio is related to excess air by the following equation:

$$\phi = 0.7242 - 0.0022 \text{ (EA percent)} \tag{3-2}$$

Then, by substitution, excess air can be related directly to
adiabatic flame temperature (°F) by the following expression:

$$T_{ad} = 2794 - 8.224 \text{(EA percent)} \tag{3-3}$$

In summary, every percentage increase in excess air decreases
the adiabatic flame temperature in the system by a little over
8°F (4.5°C).

The impact of excess air on efficiency is equally significant. Again using the Hogdalen boiler parameters with the
exception of excess air (particularly using the Hogdalen final
stack temperature of 365°F), an approximation equation was
developed for efficiency as shown below:

$$\eta = 76.87 - 0.057 \text{ (EA percent)} \tag{3-4}$$

Alternatively, for every 10% increase in excess air, there is
a decrease in boiler thermal efficiency of about 0.6%. Given
the range of excess air levels acceptable in MSW mass burning,
these efficiency losses merit control.

b. *Influences of Excess Air on Airborne Emissions.* The
amount of excess air provided in an incinerator influences not
only flame temperature and thermal efficiency, but also
organic emissions such as carbon monoxide (CO), hydrocarbons
(HCs), and dioxins and furans. Excess air can influence the
formation of particulates and oxides of nitrogen (NO_x) as
well.

The organic emission rates and formation mechanisms are
all interconnected. At very low levels of excess air, insufficient oxygen is available to the radicals and fragments in
the firebox, and CO and HC emissions rise. Essentially the
firebox is acting much like a gasifier, and portions of the
firebox are operating in a "fuel rich" mode. Equivalence
ratios approach 1.0, and may exceed 1.0 on a localized basis.
Conversely, when excess oxygen levels are very high (i.e.,
equivalence ratios are less than 0.45), the temperature in
the firebox is depressed. Since CO combustion is limited by

kinetics, incomplete oxidation of CO occurs. Again, organic
emissions rise. The importance of excess air to CO emissions
is shown in Chapter II.

Excess air influences the formation of dioxins and furans,
particularly the tetrachlorinated isomers of these classes of
compounds (Table V) as indicated by the data base developed
by Beychok [4]. This data base demonstrates that the lowest
dioxin and furan emissions are generally associated with
incinerators using 80-140% excess air. The Beychok data are
particularly significant because they have been normalized to
a common analytical framework. Further, they offer limited
statistical insights into the role of excess air in dioxin
formation. Specifically, for mass burn incinerators using
greater than 80% excess air, dioxin and furan emissions can
be approximated by the following equation:

$$D+F_{ug}/nm^3 \text{(adjusted to 12 percent } CO_2 \text{, dry basis)} =$$
$$0.0376 \text{(EA percent)} - 3.305 \qquad (3-5)$$

The coefficient of correlation (r) for equation 3-5 is 0.944
The coefficient of determination (r^2) for equation 3-5 is
0.891.

Equation 3-5 is statistically interesting, although not
predictive in any sense [4]. It demonstrates the general
correlation between high levels of excess air and high levels
of dioxin and furan formation. As Chapter II demonstrates,
however, mechanisms associated with the formation of dioxins
are highly complex, and are a direct function of waste composition and temperature, as well as oxygen availability. Consequently, the influence of excess air on dioxin and furan
emissions is probably through the control and dampening of
combustion temperature, rather than through any direct formation mechanism.

Excess air has only a minor role in the formation of particulate and NO_x emissions from the mass burning of mixed MSW.
The influence of excess air on particulate emissions is
limited to two phenomena: (1) high levels of excess air entrain
more solid particles in the gas stream, leading to increased
grain loadings of particulates influent to the AQCS; and
(2) high levels of excess air increase the velocity of gaseous
flow through the firebox, decreasing the residence time of
combustibles in the furnace and thereby decreasing the completeness of combustion. This latter phenomenon is comple-
mented by the lower flame temperatures, which in turn lead to
longer required residence times for complete oxidation of the
fuel. The influence of excess air in the formation of NO_x
from the combustion of MSW is of relatively minor consequence.
It is limited by the relatively low combustion temperatures

TABLE V. Data Base for Dioxin and Furan Emissions
from Municipal Solid Waste Incinerators

Plant location	Average μg NM^3, dry 12% CO_2, ASR[a]			Average μg NM^3, dry 12% CO_2, USR[b]		
	D_{4-8}	F_{4-8}	D+F	D_{4-8}	F_{4-8}	D+F
Hamilton, Can[d]	24.10	57.20	81.30	N/A	N/A	N/A
Hamilton, Can[d]	4.40	9.40	13.80	2.50	4.93	7.43
Toronto, Can	3.55	3.49	7.04	2.11	3.09	5.20
Chicago, IL	0.15	0.62	0.77	0.06	0.34	0.40
Albany, NY	0.95	0.21	1.16	0.51	0.11	0.63
Hampton, VA	2.80	10.56	13.36	2.32	8.56	10.78
Tsushima, Japan	2.04	2.48	4.52	N/A	N/A	N/A
P.E.I., Can	0.11	0.14	0.25	0.09	0.11	0.20
Hamburg I, Ger[e]	0.43	0.31	0.74	0.26	0.21	0.47
Hamburg II, Ger[e]	0.21	0.42	0.63	0.13	0.28	0.41
Hamburg III, Ger[e]	0.10	0.18	0.28	0.06	0.12	0.18
Como, Italy	2.65	2.91	5.56	2.25	2.47	4.72
Zurich, Switz.	0.21	0.15	0.36	0.13	0.10	0.23
Zaanstad, Hol.	2.51	2.88	5.39	2.01	2.30	4.31
Stuttgart, Ger.	0.39	1.02	1.41	0.20	0.52	0.72
Harelbeke, Belg.	7.64	8.15	15.79	4.66	5.47	10.13
Brasschaat, Belg.	1.62	6.00	7.62	0.99	4.03	5.02
Averages[f]	1.36	3.06	4.92			

[a]Emissions of tetra through octachlorinated dioxins and furans, in μg Nm^{-3} of dry stack gas at 12% CO_2, adjusted for spike recovery (ASR).

[b]Emissions of tetra through octachlorinated dioxins and furans, in μg Nm^{-3} of dry stack gas at 12% CO_2, unadjusted for spike recovery (USR).

[c]Nominal MSW feed capacity per each incinerator and type of MSW.

TABLE V. (Continued)

Number of tests	Test year	MSW burned[c]		Average excess air (%)	Steam boiler	Partic. control	Incin- erator manu- facturer
		Type	MgD^{-1}				
3	1982	RDF	272	286	yes	ESP	B and W
13	1983	RDF	272	131	yes	ESP	B and W
3	1981	Mass	192	348	no	ESP	Detroit Stoker
3	1918	Mass	363	85	yes	ESP	Martin
6	1984	RDF	273	94	yes	ESP	Zurn
5	1983	Mass	114	50	yes	ESP	Detroit Stoker
2	1983	Mass	150	197	no	S+B	Martin
12	1984	Mass	33	137	yes	none	Consumat
2	1984	Mass	384	165	yes	ESP+S	Martin
2	1984	Mass	156	79	yes	ESP+S	Martin
3	1984	Mass	504	121	yes	ESP+S	Widmar- Ernst
17	1981	Mass	85	250	yes	ESP+S	Bartolo- meis
1	1981	Mass	400	88	yes	ESP	Martin
8	1979	Mass	195	180	no	ESP	Peters
16	1985	Mass	480	105	yes	ESP	Martin- Babcock
21	1983	Mass	132	456	no	ESP	C.E.C.
32	1983	Mass	60	268	no	ESP	Volund

[d] The Hamilton MSW burning rate was reduced by 20% and the ESP was modified between the 1982 and the 1983 test programs.

[e] The Hamburg I, II, and III units are known as the Borsig-trasse, Stellinger Moor and Staplefield units, respectively.

[f] The averages doe not include the 1982 test data from Hamilton, Canada.

associated with mass burning of MSW, and the consequent
absence of thermal NO_x (see Chapter II for discussion of the
mechanisms of NO_x formation). Excess air can increase NO_x
formation from fuel nitrogen, however. As pershing and coworkers have shown [14], increased availability of oxygen
within the furnace can increase the oxidation of nitrogen in
combustion systems. Given the requirement for some 80% excess
air, at a minimum, in the mass burning of MSW: changes from
80 to 100 or 110% excess air will have minimal consequence.

2. Combustion Air Distribution

Control of the total quantity of air used in the combustion process is one of two major air-related variables. The
second issue of significance is the distribution of that air.
There are two air distribution questions: (1) distribution
of air associated with the grate system, and (2) distribution
of air in the furnace as a whole, with emphasis on the under-
fire/overfire air ratios. Distribution of air under the grate
requires sufficient adjustable plenums to ensure adequate air
to all portions of the grate. Sufficient distribution provides for proper control of all gas-solid phases of the combustion mechanism including heating and drying, solid particle
pyrolysis, and char oxidation (see Chapter II). Most grates
have at least four separate and independently controlled air
plenums (see, for example, Fig. 3 for the air distribution in
the Hogdalen installation).

The additional air distribution issue is the extent of
overfire air utilization. Typically in mass burn installations, air is introduced both as primary (undergrate) air and
as secondary (overfire) air. The distribution of such air
influences the formation of organic emissions such as dioxins
and furans, and the control of oxides of nitrogen through
staged combustion.

Combustion air distribution may be expressed either as
percentage of total air being used as overfire air, or it may
be expressed in terms of grate stoichiometric ratio (GSR).
The latter is calculated as follows:

$$GSR = (\text{lb-m } O_2, \text{ primary air} + \text{lb-m } O_2, \text{ fuel})/\text{lb-m } O_2$$

required for complete oxidation of the carbon,

hydrogen, and sulfur contained in the fuel.   (3-6)

Staged combustion occurs when GSR<1.0. Unstaged combustion occurs when GSR>1.0. In all cases, the method of secondary air injection is quite significant. Air nozzle design to

maximize penetration and turbulence is desired. Similarly,
proper placement of secondary air nozzles is necessary to
optimize the combustion process.

    *a. Influence of Combustion Air Distribution on System
Performance.* The influence of combustion air, and whether
or not staging is utilized, can have significant influence on
system performance. Data available [5] indicate that ash
fusion temperatures for municipal waste are relatively low,
particularly under reducing atmosphere conditions. Such
temperatures are shown in Table VI. Promoting reducing conditions in the combustion chamber has the consequence of
increasing metal tube wastage if refractory is not used,
particularly in the lower portions of the furnace. This condition was experienced by Nashville Thermal Transfer Corp.
on its boilers No. 1 and No. 2. The consequence is increased
maintenance and reduced unit availability.

    Combustion air distribution, and the consequent potential
for slagging, also can impact refractory selection and refractory life. Increased potential for slagging imposes requirements for more costly refractory installations. System performance requirements generally illustrate the utility of
avoiding staging conditions to the greatest extent possible.

TABLE VI. Ash Fusion Temperatures

	Reducing atmosphere		Oxidizing atmosphere	
	°F	°C	°F	°C
Refuse				
Initial Def.	1880–2060	1030–1130	2030–2100	1110–1150
Softening	2190–2370	1200–1300	2260–2410	1240–1320
Fluid	2400–2560	1315–1400	2480–2700	1360–1480
Coal				
Initial Def.	1940–2010	1060–1100	2020–2270	1100–1240
Softening	1980–2200	1080–1200	2120–2450	1160–1340
Fluid	2250–2600	1230–1430	2390–2610	1310–1430

b. Influence of Combustion Air Distribution on Airborne Emissions. The influence of air distribution on airborne emissions includes considerations of organic emissions (i.e., dioxins and furans), particulates, and NO_x. Combustion air plays a substantial role in controlling the formation of dioxins and furans as has been suggested by numerous authors. The literature generally support the concept that the avoidance of staged combustion, coupled with high temperatures, minimizes the formation of dioxins and furans from mass burning of MSW [17, 18]. Mathematically the avoidance of staging can be estimated by the following equation derived specifically for mass burning installations:

$$GSR_{mb} = 1.511 + (0.112 \times UFA/OFA) - 1.175(O) \qquad (3\text{-}7)$$

Where UFA/OFA is the ratio of the percentage of underfire air to overfire air. When 40% of the total air is supplied as overfire air, the UFA/OFA term is 1.5.

Typically, mass burn plants use 30-40% of total air as overfire air. Consequently, the UFA/OFA term is 1.5 - 2.33. Typical GSR_{mb} values for well run incinerators with excess air levels of 80-110% (equivalence ratios of 0.556 - 0.476) have grate stoichiometric ratios of 1.02 - 1.21.

The influence of air distribution on particulate emissions is of some importance. Currently the typical mass burn installations have a bottom ash/collected flyash production ratio of about 3:1; some 75% of the solid products of combustion, including inerts and unburned carbon, are removed from the system as bottom ash while only about 25% of the solid products are collected as flyash. Bottom ash is removed from the toe of the grate while collected flyash is removed from the knockout box in the economizer and from the AQCS system. Increasing the use of overfire air, and simultaneously decreasing the use of undergrate air, will decrease the percentage of solid products reporting as flyash. The consequence of such actions is decreased particulate grain loadings influent to the AQCS [12]. This combustion air manipulation can be pushed to the extreme, however, compromising the control of organic emissions as discussed above.

The minimization of NO_x formation by increasing the use of overfire air and consequently using staged combustion (achieving a GSR<1.0) also conflicts with the minimization of dioxins and furans as discussed above. Munro [12] demonstrated that NO_x formation from conversion of the fuel nitrogen in biomass fuels is minimized with grate stoichiometric ratios of about 0.8. However the use of such GSR values implies the need for very low levels of excess air (i.e., 40% excess air) and very high levels of overfire air (i.e., 60% OFA). The control air

distribution to minimize NO_x formation is not particularly
productive for the mass burning of MSW when the total range of
emissions is considered, and when the availability of post-
combustion controls for NO_x is recognized (see Chapter VIII).

3. Combustion Air Temperature

The temperature of the combustion air or, specifically,
the preheating of combustion air, is also significant. Preheating of combustion air introduces more enthalpy into the
furnace, and consequently raises the flame temperature. It
has been shown that, for biomass and waste type fuels, every
degree of air preheat above 77°F (25°C) increases the flame
temperature by 0.6°F (0.3°C) [23].
Typically, combustion air is not preheated in MSW mass
burning installations. Provisions are made for preheating,
however, in the cases of unusually wet MSW being fed to the
incinerator. Under such circumstances preheated air provides
for feedstock drying and temperature maintenance, the latter
being necessary for control of dioxins and furans. When
combustion air is preheated, the distribution of preheat may
favor underfire air (for grate drying) or overfire air (for
increased turbulence and penetration). The choice is based
upon site specific and design specific factors.

4. Combustion Air Conclusions

The manipulation of combustion air is essential to the
process of mass burning. Several issues exist including the
level of excess air, the distribution of combustion air, and
the temperature of combustion air. Good combustion practices
for mass burning of MSW have been developed by EER Corporation
[17] and most of the considerations relate directly or indirectly to the control of combustion air (Table VII).

B. Rates of Heat Release

The second combustion control issue of significance
involves the rates of heat release associated with mass burn
installations, including grate heat release and volumetric
heat release. These design rates chosen determine the shape
and size of the furnace, and many of the parameters associated
with combustion residence time and gaseous turbulence. Further they impact the mass flow of waste through the incinerator (throughput rate), and consequently the capital utilization effectiveness of the system. High rates of heat release,
both in terms of grate heat release and volumetric heat

TABLE VII. Good Combustion Practices for Minimizing
Trace Organic Emissions from Mass Burn Municipal
Waste Combustors

Element	Component	Recommendations
Desian	Temperature at fully mixed height	1800°F (982°C) at fully mixed height
	Underfire air control	At least four separately adjustable plenums. One each under the drying and burnout zones and at least two separately adjustable plenums under the burning zone
	Overfire air capacity (not an operating requirement)	40% of total air
	Overfire air injector design	That required for penetration and coverage of furnace cross-section
	Auxiliary fuel capacity	That required to meet start-up temperature and 1800°F criteria under part-load operations
Operation/ control	Excess air	6-12% oxygen in flue gas (dry basis)
	Turndown restrictions	80-100% of design--lower limit may be extended with verification tests
	Start-up procedures	On auxiliary fuel to design temperature
Verification	Oxygen in flue gas	6-12% dry basis
	CO in flue gas	50 ppm on 4-hr average-- corrected to 12% CO_2
	Furnace temperature	Minimum of 1800°F (mean) at fully mixed height across furnace
	Adequate air distribution	Verification tests (see text Chapters 8 and 9)

Source: [17].

release, represent increased throughput in the system (lb of
waste per sq ft or per cu ft per hour) and consequently
improved capital utilization rates. One combustion process
consequence is less solids residence time on the grate and
less gaseous residence time in the furnace. Additional consequences for the combustion process include the potential
for lower rates of carbon conversion due to residence time
effects, the potential for increased particulate loadings
influent to the AQCS due to gaseous velocity effects, and the
possibility of higher combustion temperatures. There is a
practical tradeoff, however. If rates of heat release are
too low, the fire may become increasingly "lazy," and this
may reduce combustion completeness. Further, the grate may
be underloaded and subject to bare spots. Finally, very low
rates of volumetric heat release can result in combustion
with inadequate temperature.

Historically, mass burn incinerators have been designed
with a wide range of heat release rates as shown from European
experience (Table VIII). Grate heat release rates ranged from
139,000 Btu/sq ft-hr to 506,000 Btu/sq ft-hr. Similarly
volumetric heat release rates ranged from 8,500 Btu/cu ft-hr
to 30,000 Btu/cu ft-hr over this selection of systems. Mean
values from this sample of incinerators include a grate heat
release rate of about 320,000 Btu/sq ft-hr and a volumetric
heat release rate approaching 20,000 Btu/cu ft-hr. Data from
a small sample of U.S. plants demonstrates a somewhat narrower
range of heat release values (Table IX). Again these values
converge on 330,000 Btu/sq ft-hr for a grate heat release
rate. However there is a more conservative volumetric heat
release rate of about 10,000 Btu/cu ft-hr. These rates of
heat release are comparable to those shown for the Hogdalen,
Sweden, plant.

Recently, there has been considerable convergence on the
heat release rates of about 300,000-330,000 Btu/sq ft-hr on
the grate and 8,500-10,000 Btu/cu ft-hr in the furnace as
exhibited not only in constructed units but also proposed
units. Exceptions have resulted from design uniqueness (i.e.,
use of rotating kilns). Such values, then, appear to represent the consensus of compromise for combustion manipulation
between the requirements of capital utilization and the
requirements of combustion completeness and thermochemistry
control.

C. Combustion Process Calculations

    Control of combustion air and rates of heat release for
the purposes of minimizing emissions and maximizing efficiency
are the fundamental considerations associated with mass

TABLE VIII. Heat Rlease Rates for Selected European
Mass Burn Systems

System	HHV Basis		LHV Basis	
	Btu/ft^2 grate per hr	Btu/ft^3 furnace per hr	Btu/ft^3 grate per hr	Btu/ft^3 furnace per hr
Wardenberg – Lichtenstein	302,642	11,945	280,224	11,060
Baden–Brugg	275,970	15,876	255,528	14,700
Dusseldorf				
Boilers 1-4	190,404	12,582	176,300	11,650
Boiler 5	278,720	11,951	259,000	11,066
Wuppertal	376,380	24,829	348,500	22,990
Krefeld	301,118	119,167	278,813	17,747
Paris, Issy	229,897	22,717	212,868	21,034
Hamburg	305,480	24,686	282,852	22,857
	334,174	27,063	309,420	25,058
The Hague				
Boilers 1-3	340,472	14,096	315,252	13,052
Boiler 4	336,234	16,197	311,328	14,997
Dieppe	139,244	8,537	128,930	7,905
Goteberg	335,340	14,391	310,500	13,325
Uppsala	300,470	29,694	278,212	27,494[a]
Horsens	343,466	27,365	318,024	25,338
Copenhagen Amager	505,566	–	468,117	–
Copenhagen West	505,566	–	468,117	–
Mean	317,713	18,740	294,234	17,352
Standard deviation	92,234	6,769	85,375	6,268
Range				
Low	139,244–	8,537–	128,930–	7,905–
High	505,566	29,694	468,117	27,494

[a]Port of combustion occurs in cyclonic of afterburner.
Source: [2].

TABLE IX. Heat Release Rates for Selected
North American Plants

| Plant | Design heat release rate | |
	Btu/ft^2 grate per hr	Btu/ft^3 furnace per hr [a]
Nashville		
Units 1 and 2	217,740	10,100
Unit 4	292,500	10,100
Pinellas County[b]	410,000	9,100
North Andover[b]	360,000	10,100
Baltimore	328,180	N/A
Tampa	250,000	14,000
Gallatin	N/A	6,000
Mean	328,200	N/C[a]

[a]Proposal values the authors have seen range from 6,500–25,000 Btu/ft . This convergence in the sample is coincidental.

[b]Grate sizes are identical (∿800 ft^2/boiler). Differences are in Btu content of design fuel.

Source: Plant visits [24].

burning of MSW. Generally the following conditions appear to be the consensus of current practice: (1) excess air – 80–110%; (2) air distribution – GSR>1.0, overfire air up to 40%; (3) consequent combustion temperature – >1800°F for 1–2 sec.; (4) combustion air temperature – preheated to 300°F only when required for combustion temperature maintenance due to unusually wet fuel; (5) grate heat release – 300,000–330,000 Btu/sq ft-hr; and (6) volumetric heat release – 8,500–10,000 Btu/cu ft hr. These may be varied if required by an unusual system design.

IV. MASS BURN SYSTEM COSTS

The process of mass burning MSW carries the potential for
mass and volume reduction of materials being sent to landfill.
Costs of consequence include both capital and operating costs.
These, along with local conditions such as ownership and
regulatory requirements, can be converted into specific tipping fees.

A. Capital Costs

Analytically there are several approaches to establishing
the capital costs of mass burn facilities including developing
capital cost estimates for specific pieces of equipment and
aggregating these into total system capital costs, evaluating
historical capital costs, and evaluating recent proposals.
Given the fact that developing capital costs from specific
pieces of equipment and subsystems is highly project-specific,
that approach is not taken here. Rather, evaluation of historical costs and evaluation of recent total proposal costs
is used for this analysis.

1. Historical Capital Costs of Mass Burn Facilities

Historical data on capital costs have been published by
Waste Age (see, for example, Smith [22]), by the Resource
Recovery Yearbook, and by other sources. The data published
by Waste Age indicate that, historically, mass burn facilities
have cost anywhere from $42,000/daily ton of capacity to
$92,000/daily ton of capacity when the capital charges are
expressed in 1988 dollars (Table X). It is significant to
note that the facilities with capacities less than 1,00 TPD
have had costs ranging from about $56,000 to $92,000/daily
ton of capacity while facilities with capacities greater than
1,000 TPD have had costs ranging from $42,000 to $82,000/
daily ton of capacity. These data do not lend themselves to
detailed and rigorous statistical analysis. Generally, however, two regression equations for cost approximation emerge:

$$C_{act} \text{ (in } \$\times10^6) = .066T + 4.184Y - 9620 \tag{3-8}$$

and

$$C_{1988} \text{ (in } \$\times10^6) + .066T + 1.848Y - 3662 \tag{3-9}$$

TABLE X. Capital Costs of Representative MSW Mass Burn Plants

Plant location	Size (TPD)	Year installed	Capital cost actual	Capital cost ($ × 10⁶) 1988 $	Capital cost actual	($/daily ton) 1988
Pinellas County, FL	2,000[a]	1983	80.0	83.9	40,000	42,000
Tampa, FL	1,000	1985	59.9[c]	67.1	60,000	67,000
Chicago, IL (NW)	1,600	1970	23.0	76.5	74,000	48,000
Baltimore, MD	2,250	1985	170.0	173.8	76,000	77,000
N. Andover, MA	1,500	1985	196.0	200.0	131,000	134,000
Saugus, MA	1,500	1975	50.0	91.1	33,000	61,000
Claremont, NH	200	1987	17.9	18.4	90,000	92,000
Westchester, NY	2,250	1984	179.0	184.4	80,000	82,000
Tulsa, OK	750	1986	51.5	53.8	69,000	72,000
Marion County, OR	550	1986	47.5	49.6	86,000	90,000
Nashville, TN	712[a]	1974	24.5	49.3	34,000	69,000
Gallatin, TN	200	1981	10.0	11.2	50,000	56,000
Montreal, Canada	1,200	1970	14.7[b]	58.7[d]	12,000	49,000
Quebec, Canada	1,000	1974	25.0[b]	60.3[d]	25,000	60,000

[a]Original size. [b]Canadian dollars. [c]1981 dollars. [d]Converted to U.S. dollars at
80% rate. Sources: [22], CE Plant Cost Index (332.5 = Jan. 1, 1988).

Where T is tons per day (TPD) of capacity and Y is calendar
year. These equations were developed from costs for the
projects ranging in size from 550 TPD to 2,250 TPD. The coefficients of determination (r^2) for these equations are 0.83
for equation 3-8 and 0.80 for equation 3-9. These data generally show that the economies of scale are minimal for
field erected mass burn units. These equations also show
that there has been a significant real cost increase in mass
burn incineration systems over time, as regulatory requirements become more stringent and as development risks are perceived to increase.

It is useful to compare these data to studies by others.
Kerstetter and Schlorff [10] have shown that mass burn units
generating electricity have a capital cost of $39,300–
$111,800/daily ton of capacity (in 1985 dollars), with a
median cost of $71,670/daily ton of capacity. These data,
converted to 1988 dollars by the CE Plant index, yield a
median value of $73,300/daily ton of capacity for mass burn-
electricity generating plants. Kerstetter and Schlorff also
show no particular economies of scale with field erected mass
burn installations [10]. Data developed by Brunner [5] similarly show the absence of economies of scale in large scale
mass burn installations.

2. *Current and Projected Capital Costs of Mass Burn
   Facilities*

Historical data do not reflect many of the regulatory
trends currently impacting development risk and installation
requirements associated with mass burn installations. Such
risks include public opposition based upon perceived organic
emissions such as dioxins and furans, post-combustion controls
for NO_x control (i.e., ammonia injection), and other concerns.
Consequently, a step-function increase in capital costs has
been observed for mass burn installations.

An analysis by the authors of sixteen proposed mass burn
facilities identified in Waste to Energy Reporter yielded a
capital cost function as follows:

$$C_{proposed} = 16.23 + 0.086T \qquad (3\text{-}10)$$

Where $C_{proposed}$ is the capital cost of the proposed facilities
(in millions of 1988 dollars), and T is daily tons of capacity.
The coefficient of determination (r^2) for this equation is
0.725. This equation provided the best fit for the data.
Equation 3-10 shows that there is generally a minimum cost,
or "price of admission," for mass burn facilities in excess

of $16 million. Beyond amortization of that price, there
are few economies of scale.

The data for proposed facilities can be compared to the
results of the Kerstetter and Schlorff [10] survey. For systems in the 1,000 to 3,000 TPD range, the projected facilities
will have a capital cost of about $91,000 to $102,000/daily
ton of capacity. This represents a significant increase from
the historically experienced median of $73,000/daily ton of
capacity (in 1988 dollars) as shown by those researchers.
Clearly the capital cost of facilities designed for complete
manipulation of the combustion process, for complete post-
combustion control of the airborne emissions, and for development risk accommodation has responded to recent changes in the
socio-political climate.

B. Operating Costs

Like capital costs, operating costs indicate some initial
"price of admission," and then few economies of scale. Operating and maintenance costs have been analyzed both in terms of
labor requirements and in terms of total costs as shown below.

Operating labor and total labor dominate the operating
costs associated with a mass burn facility. A survey taken
by the authors, using total facility labor as a proxy for
operating cost, yielded the following approximation regression
equation:

$$Lt = 24.22 + 0.033T \qquad\qquad (3-11)$$

The correlation coefficient (r) for this equation is 0.83.
The only economies of scale found with respect to labor
requirements are associated with recovering the cost of the
initial 24 workers over a broader tonnage range.

Kerstetter and Schlorff [10] found total operating and
maintenance costs for mass burn facilities to range from
$16.50/ton to $32.80/ton in 1985 dollars, or $17/ton to $34
in 1988 dollars. The median value cited by these researchers
was $24.10/ton in 1985 dollars, or $25/ton in 1988 dollars.
Operating and maintenance costs are also expected to increase
for future installations as more complex post-combustion controls are added to systems, and as ash disposal becomes of
increasing concern.

C. Cost Conclusions

It is difficult to generalize about disposal costs given
the widely ranging regulatory climates in the U.S., the ownership differences, the financing methods used, and other key
variables. It can be concluded, however, that future installations of well designed mass burn facilities can be constructed and operated for cost-effective waste disposal. Such
costs will be impacted by increased societal attention to the
processes of combustion as they impact the environment.

V. OVERALL CONCLUSIONS

Mass burning of MSW has become a well developed, proven
method for waste management and disposal. Facilities throughout the U.S. and Europe have demonstrated that this technology
can be applied in an environmentally sound manner. Further,
the process of combustion can be controlled to achieve efficiency, environmental, and economic benefit. In order to
achieve such objectives, however, systems have become
increasingly expensive to construct and operate. However,
mass burn installations can be developed within the economic
constraints of a cost-effective solution to the waste management problem.

REFERENCES

1.   Anon. 1985. Nashville, Tennessee: A Waste-to-Energy
     Case Study. United States Conference of Mayors,
     Washington, D.C.

2.   Battelle Columbus Laboratories. 1979. European Refuse
     Fired Energy Systems: Evaluation of Design Practices.
     U.S. Department of Commerce, National Technical Informa-
     tion Service, Washington, D.C. (for U.S.-E.P.A.).

3.   Berenyi, E., and Gould, R. (eds.). 1986. 1986-87
     Resource Recovery Yearbook. Government Advisory Asso-
     ciates, Inc., New York.

4.   Beychok, M. R. 1987. A Data Base of Dioxin and Furan
     Emissions from Municipal Refuse Incinerators. *Atmos-
     pheric environment 21*(1):29-36.

5. Brunner, C. R. 1986. Mass Burn Incineration: State-of-the-Art Review. Presented at the Specialty Conference on Incineration of Municipal Waste, Bellevue, Washington (CH2M Hill).

6. Cooper Engineers, Inc. 1984. Air Emissions Tests of Solid Waste Combustion in a Rotary Combustor/Boiler System at Gallatin, Tennessee. Cooper Engineers, Richmond, California.

7. Hasselriis, F. 1986. Effects of Burning Municipal Waste on Environment and Health. Presented at the American Society of Mechanical Engineers/Institute of Electrical and Electronic Engineers Joint Power Generation Conference, Portland, Oregon. Paper #: 86-JPGC-EC-17.

8. Hazard, N. D., and Tease, H. V. 1986. Factors in the Selection of Boilers for Firing Refuse Fuels. Presented at the Arizona Water and Pollution Control Association Annual Conference, Phoenix, Arizona. (Combustion Engineering Paper #: RRS-1002).

9. Johnson, N. H., and Reschly, D. C. 1986. MSW and RDF - An Examination of the Combustion Process. Paper presented at the American Society of Mechanical Engineers/Institute of Electrical and Electronic Engineers Joint Power Generation Conference, Portland, Oregon. Paper #: 86-JPGC-Pwr-20.

10. Kerstetter, J., and Schlorff, E. 1987. Municipal Solid Waste to Energy: Analysis of a National Survey. Washington State Energy Office, Olympia, Washington.

11. McNertney, R. M., Chambliss, C. W., and Hestle, J. T. 1986. Operating Experience is Reflected in Nashville's New Refuse Boiler. Paper presented at the American society of Mechanical Engineers/Institute of Electrical and Electronic Engineers Joint Power Generation Conference, Portland, Oregon. Paper #: 86-JPGC-Pwr-19.

12. Munro, J. 1982. Formation and Control of Pollutant Emissions in Spreader-Stoker-Fired Furnaces. PhD Dissertation. University of Utah, Salt Lake City, Utah.

13. Parker, C. et al. 1985. Sumner County Solid Waste Energy Recovery Facility, Vol 1: Feasibility Studies, Design, and Construction. Final Report Prepared by the Tennessee Valley Authority, Chattanooga, Tennessee. Electric Power Research Institute, Palo Alto, California.

14. Pershing, D. W., and Wendt, J. 1976. Pulvarized Coal Combustion: The Influence of Flame Temperature and Coal Composition on Thermal and Fuel NO_x. Proceedings of the Sixteenth International Symposium on Combustion, The Combustion Institute.

15. Resource Consultants, Inc. 1983. Particulate, SO_2, NO_x, and CO Emissions Source Testing Program Conducted bon Municipal Waste Incinerator Boiler at Nashville Thermal Transfer Corporation, Nashville, Tennessee. Resource Consultants, Inc., Brentwood, Tennessee.

16. Scherrer, R., and Juran, D. 1986. Air Pollution Control Considerations for Municipal Solid Waste-to Energy Facilities. Presented at the American Society of Mechanical Engineers/Institute of Electrical and Electronic Engineers Joint Power Conference, Portland, Oregon. Paper #: JPGC-EC-12.

17. Seeker, W. R., Lanier, W. S., and Heap, M. P. 1987. Combustion Control of MSW Incinerators to Minimize Emissions of Trace Organics. EER Corporation, Irvine, California. For USEPA Contract No. 68-02-4247.

18. Shaub, W. M., and Tsang, W. 1983. Dioxin Formation in Incinerators. *Environmental Science and Technology 17* (12:721-730.

19. Signal Environmental Systems, Inc. 1985. Baltimore's Refuse-to-Energy Facility: Burning Garbage Downtown. Advertising Section in *Waste Age* Magazine.

20. Signal Environmental Systems, Inc. (Undated.) Baltimore, Maryland Refuse-to-Energy Facility (fact sheet).

21. Signal Environmental Systems, Inc. (Undated.) Pinellas County, Florida Refuse-to-Energy Facility (fact sheet).

22. Smith, J. A. 1985. Energy and Materials Recovery Facilities. *Waste Age 16*(11:99-138.

23. Tillman, D. A., and Anderson, L. L. 1983. Computer Modelling of Wood Combustion with Emphasis on Adiabatic Flame Temperature. Journal of Applied Polymer Science: Applied Polymer Symposium 37:761-774.

24. Tillman, D. A. 1987. Personal communication. Plant visits and tours of the following facilities: Tampa (McKay Bay), Pinellas County, Florida; Baltimore, Maryland;, North Andover, Massachusetts; Gallatin, Tennessee; Nashville Thermal Transfer; Hogdalen, Sweden; Malmo, Sweden; and Gothenberg, Sweden.

25. Turner, W. D. 1982. Thermal Systems for Conversion of Municipal Solid Waste, Vol. 2: Mass Burning of Solid Waste in Large-Scale Combustors: A Technology Status Report. Argonne National Laboratory, Argonne, Illinois

26. Waste Management, Inc. (with City of Tampa). 1985. McKay Bay Refuse-to-Energy Facility, Tampa, Florida (fact sheet and project description).

27. Waste-to-Energy Report, June 3, 1987.

Chapter IV

THE PRODUCTION AND COMBUSTION OF

REFUSE DERIVED FUELS

I. INTRODUCTION

Refuse derived fuel (RDF) systems offer a practical, commercially proven alternative to mass burning of municipal waste. These systems manipulate both the fuel properties and the combustion system. RDF projects attempt to produce a more uniform "fuel" to be burned, and consequently to provide tighter control and more efficient combustion. RDF technology processes raw refuse prior to burning the material. Processing steps are designed to achieve particle size reduction and to remove fractions of waste which are high in noncombustible materials (i.e., ferrous metal and glass) or high in moisture content (i.e., yard trimmings). The resulting product is a more useful fuel which facilitates the combustion process. RDF technology was originated in the 1970s. Since commercial introduction over 20 RDF units have been constructed in the U.S. as shown in Table I. Additional units have been constructed throughout the industrialized world.

A. RDF Technology Overview

Distinct fuel production and combustion technologies exist as part of the RDF alternative. Production technologies focus on the types of RDF produced. Fuel types as produced are formally defined by the American Society for Testing and Materials, as shown in Table II. As a practical matter, there are three grades of RDF for consideration here: (1) coarse RDF (RDF-2 as defined by ASTM Committee E-38), (2) moderate RDF, and (3) refined or fluff RDF (RDF-3 as defined by ASTM Committee E-38). What is considered as moderate RDF here may be classified with refined or fluff RDF [27]. However, the dissimilarities between the conventional "fluff" RDF and the current generation of systems such as the Combustion Engineering (CE) plant in Hartford, Connecticut, warrant a separate classification. Typical partial fuel specifications for various grades of RDF are shown in Table III.

TABLE I.   Representative Refuse Derived Fuel Projects
in the United States

City	Capacity (TPD)	Start-up date (Year)
Akron, Ohio	1,000	1979
Ames, Iowa	200	1975
Baltimore County, Maryland	1,000	1978
Duluth, Minnesota	400	1980
Niagara Falls, New York	2,000	1981
Dade County, Florida	3,000	1982
Haverhill, Massachusetts	1,300	1984
Madison, Wisconsin	400	1979
Albany, New York	700	1981
Columbus, Ohio	2,000	1983
Hartford, Connecticut	2,000	1988
Portsmouth, Virginia	2,000	1986

Source: Smith [29].

1. RDF Production Technologies

A wide variety of separation technologies have been developed for MSW processing into RDF including: (1) trommel separation, (2) disc screen separation, (3) Brini screen separation, (4) air classification, (5) separation by the NRT segmented drum (AMSORT[R]), and (6) separation in a system using water, rather than air. All of these technologies except the water-based system are commercial today. Further, while RDF production systems are based upon the separation scheme, they also involve the following unit operations: (1) coarse shredding, (2) ferrous metal removal (magnetic separation), and (3) size classification.

Optional unit operations have been periodically attempted for aluminum recovery; glass recovery; heavy nonferrous metal recovery; plastics recovery by a variety of techniques, and secondary shredding. Many optional unit operations (i.e., glass recovery, heavy nonferrous metal recovery, plastics recovery) have not become commercial practice. RDF production processes assemble the desired unit operations into

TABLE II. ASTM Classification of Refuse-Derived Fuels

Type of RDF	Description
RDF-1 (MSW)	Municipal solid waste used as a fuel in as-discarded form, without oversize bulky waste (OBW).
RDF-2 (c-RDF)	MSW processed to coarse particle size, with or without ferrous metal separation, such as 95% by weight (weight % passes through a 6-inch square mesh screen.
RDF-3 (f-RDF)	Shredded fuel derived from MSW and processed for the removal of etal, glass, and other entrained inorganics. The particle size of this material is such that 95 wt % passes through a 2-inch square mesh screen. Also called refined RDF.
RDF-4 (p-RDF)	Combustible-waste fraction processed into powdered form, 95 wt % passing through a 10-inch (0.035-inch square) screen.
RDF-5 (d-RDF)	Combustible-waste fraction densified (compressed) into the form of pellets, slugs, cubettes, briquettes, or similar form.
RDF-6	Combustible-waste fraction processed into a liquid fuel (no standards developed).
RDF-7	Combustible-waste fraction processed into a gaseous fuel (no standards developed).

Source: Adapted from ASTM Committee E-38-01 on Resource Recovery, Energy.

complete process flowsheets ranging from simple coarse shred and magnetic separation systems to complex processes for complete materials separation.

2. RDF Combustion Technologies

Associated with the various RDF production technologies are alternative combustion systems, specifically: (1) spreader-stoker systems, (2) fluidized bed systems, and

TABLE III. Typical Partial Specifications for RDF Fuels

Parameter	Coarse	RDF fuel type moderate	Refined
Mass fuel recovery rate (%)	90 - 95	75 - 90	50 - 75
Energy recovery rate (%)	95 - 99	85 - 95	70 - 85
Maximum fuel size (inches)	-6 (-15.24cm)	-4 (-10.16cm)	-2 (-5.08cm)
Fuel ash content	High (e.g., greater than 20%)	Medium (e.g., 16-20%)	Low (less than 16%)

(3) suspension firing. When spreader-stoker and fluidized bed systems are used, RDF may be fired exclusively, or in combination with coal, peat, or wood. When suspension firing is used, RDF is conventionally used to supplement pulverized coal. Currently, spreader-stoker and fluidized bed systems appear to dominate the RDF combustion installations.

RDF production also has been employed as a pre-processing step for the production of alternative fuels (i.e., RDF-6 and RDF-7) as shown in Table II. Such alternative fuels include liquid petroleum substitutes and gaseous hydrocarbon substitutes [1,4]. Such energy recovery technologies never achieved commercial status, however. RDF processes also are used as the initial stages of compost production in various plants as well.

B. Organization of This Chapter

This chapter evaluates the RDF technologies from the perspective of manipulating the combustion process through control of fuel properties and application of combustion systems. It focuses on fuel production issues and analyzes alternative RDF production systems. The chapter then considers the alternative combustion systems available for RDF, focusing on spreader-stoker and fluidized bed technologies. Of particular

concern in the combustion section is the efficiency and total
energy production, given the various RDF production systems.
Also considered are airborne emissions and solid residues
associated with RDF combustion, including dioxin and furan
formation and control, and heavy metal emissions. Finally,
it examines the costs associated with producing and burning
RDF.

This chapter extensively uses case studies from existing
RDF production and combustion facilities: Columbus, Ohio;
Hartford, Connecticut; Gallatin, Tennessee; Madison, Wisconsin;
Kovik and Sundsvall, Sweden. These case studies illustrate
the different production and combustion processes available
as an alternative to mass burning of MSW.

II. RDF FUEL PRODUCTION TECHNOLOGIES

The grades of RDF identified in Table III represent points
on a continuum. That continuum is the tradeoff between fuel
quantity and fuel quality as measured by higher heating value,
moisture content, ash content, and particle size; the quality
of the fuel produced vs. the quantity of raw refuse that still
must be landfilled; and the disposition of heavy metals either
in the unburned waste or in the waste-based fuel. The production of coarse, moderate, or fluff RDF is accomplished by
assembling various unit operations into process systems.
Selection and arrangement of such unit operations governs the
mass and energy yields of fuel from the given RDF plant. The
most common unit operations are the particle size reduction
systems (shredders), "decision makers" (i.e., trommel, air
classifier), and ferrous metal recovery (magnets). Mass
yields of these unit operations are shown in Table IV. Supplementary operations may be employed for recovery of aluminum,
or other products. Aluminum recovery by eddy current or
"aluminum magnet" separation, for example, may be employed,
with mass recovery rates of about 76% [10].

Any analysis of various RDF production systems necessarily
involves an examination of combinations of these unit operations. A baseline composition of MSW was previously presented
in Chapter II, Table II, and it is used to support this analysis. Due to their importance in environmental analysis,
sources of heavy metals are presented in Table V.

A. Production of Coarse RDF

Coarse RDF involves minimal processing. Existing plants
include Columbus, Ohio; Albany, New York (Answers), and

TABLE IV. Unit Operations Efficiencies with Refuse Derived Fuel Plants

Waste material	Trommel		Air Classifier		Magnet	
	% Product (fuel)	% Reject	% Product (fuel)	% Reject	% Product (ferrous metal)	% Reject
Paper	0.95	0.05	0.595	0.405	0.005	0.995
Plastics	0.8	0.2	0.595	0.405	0	1
Rubber and leather	0.95	0.05	0.211	0.789	0	1
Wood	0.95	0.05	0.211	0.789	0	1
Textiles	0.75	0.25	0.211	0.789	0	1
Yard waste	0.25	0.75	0.211	0.789	0	1
Food waste	0.25	0.75	0.211	0.789	0.001	0.999
Fines	0.5	0.5	0.211	0.789	0.01	0.99
Glass	0.51	0.49	0.211	0.789	0	1
Ferrous metal	0.4	0.6	0.069	0.931	0.9	0.1
Aluminum	0.25	0.75	0.178	0.822	0.007	0.993
Other non-ferrous metal	0.2	0.8	0.211	0.789	0	1

Sources: Ebasco [6]; Barton [3]; and Tillman [37].

TABLE V. Percentage Distribution of Metal Content in Different Household Waste Fractions (Percentages of Each Fraction in the Waste Stream Shown in Parentheses)

Fraction	Cd	Co	Cr	Cu	Hg	Mn	Ni	Pb	Zn
Plastics (7.0)	26	1	5	2	10	1	1	5	1
Paper (46.0)	4	5	7	11	13	18	3	3	11
Animal and vegetable matter (19.0)	3	2	3	4	8	5	4	3	5
Textiles (2.0)	1	1	1	2	4	1	1	1	1
Rubber and leather (1.0)	4	1	42	1	3	1	1	2	9
Metals (6.5)	60	88	43	22	60	74	87	85	68
Miscellaneous (19.5)	3	4	3	63	3	4	6	4	6

Sources: National Swedish Environmental Protection Board [22]; also [33]. Ebasco [6]; also Mitre [19].

Niagara Falls, New York (Hooker Chemical). Typically such
plants recover 93-95% of the incoming mass of refuse as fuel,
and typically such fuel contains 97-99% of the Btu content of
the incoming refuse. Presented below is the case study of
Columbus, Ohio. Following that case study, certain generalizations can be made.

1. *The Columbus, Ohio, RDF Plant*

The coarse RDF production process installed at Columbus,
Ohio (see Fig. 1) involves shredding of 1,260 tons (1,140
tonne/day) per day (TPD) of raw refuse in two process lines.
The refuse is shredded to a -6 inches (-15 cm) nominal particle size. Following shredding the refuse is passed under a

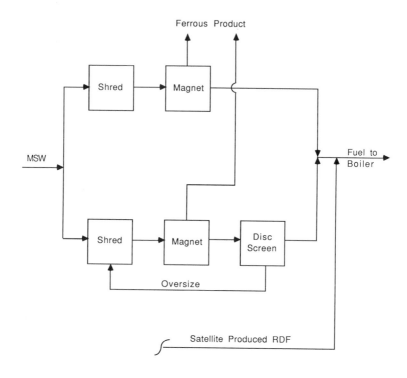

*FIGURE 1. Summary process flowsheet of the Columbus,
Ohio, RDF production plant.*

magnet where ferrous metals are removed. As shown in Fig. 1,
Columbus, Ohio, installed a disc screen processing of the
shredded waste as a means for fuel particle size control and
subsequent combustion improvement [20]. Also, as shown in
Fig. 1, the Columbus plant imports some 400 TPD of pre-
prepared RDF for use in the power plant.

2. *Coarse RDF Production Process Analysis*

The value of the coarse RDF production processes such as
the one installed at Columbus, Ohio, is not in significant
upgrading of the elemental composition of the waste stream,
as documented in Table VI. Further, the coarse RDF process as
employed in Columbus does not significantly address the heavy
metals content of the fuel being burned (see Table V).

TABLE VI. Comparison of Raw Refuse and Coarse RDF
Ultimate Analysis Based Upon Reference Fuel

Element/Compound	Wt percent by fuel	
	Raw refuse	Coarse RDF
Carbon	27.5	28.4
Hydrogen	3.7	3.8
Oxygen	20.6	21.8
Nitrogen	0.44	0.46
Chlorine	0.50	0.51
Sulfur	0.83	0.86
Inerts	23.4	20.3
Moisture	23.2	24.4
Total	100.2	100.5
Higher heating value		
Btu/lb	4830	5000
Kcal/g	2.68	2.78
Mass yield (%)	100.0	95.0
Energy yield (%)	100.0	98.3

Alternatively, however, the coarse RDF process does not increase the waste flow being directed from RDF production to
the landfill, as all of the material is either recovered as
fuel or as ferrous metal. The value of the coarse RDF production process, then, is in permitting a change in the combustion
system employed from the mass burn incinerator to a spreader-
stoker system.

B. Moderate RDF Fuel Production Processes

Moderate RDF production processes recover 75-90% of the
mass of incoming refuse as fuel, and about 85-95% of the
incoming heat content of the waste stream. Unlike coarse RDF
production processes, there are several moderate RDF systems
including trommel based systems and NRT segmented drum
(AMSORT^R) based system.

1. Case Studies in Moderate RDF Production

In order to consider these different approaches to RDF
production, two case studies are employed: (1) Hartford,
Connecticut; and (2) Gallatin, Tennessee. These case studies
highlight differences between technology and engineering
approaches to the upgrading of municipal waste based fuels.
Overall conclusions concerning moderate RDF production are presented after consideration of these specific case studies.

a. Hartford, Connecticut. The Hartford (Mid-Connecticut)
plant built by Combustion Engineering (CE) Company is a 2,000
TPD waste-to-energy facility. This plant, described by
Mirolli, Ferguson, and Bump [18], employs a trommel based
processing system as shown in Fig. 2. Incoming refuse is
coarsely shredded to -6 inches (-15 cm) in a flail mill. Following the flail mill, the refuse is passed under a magnet
for ferrous metals removal. The waste then goes through a
primary trommel where it is separated into a fuel fraction
and two underflow fractions. Separation is based upon hole
size in the trommel. The CE design includes two stages with
both large and small size holes or screen openings to differ-
entiate the waste streams [18]. The first underflow fraction
is residue which passes through screen openings designed to
reject materials less than 1/2 inch (1.3 cm) in size. This
material is largely noncombustible and is sent directly to
landfill. The second underflow fraction results from screen
openings designed to create an underflow stream of 1/2 to 1/4
inch (1.3-10.2 cm) diameter. This stream is passed through a
secondary trommel where overflow is recovered as fuel. The
fuel fraction recovered by the primary trommel is shredded

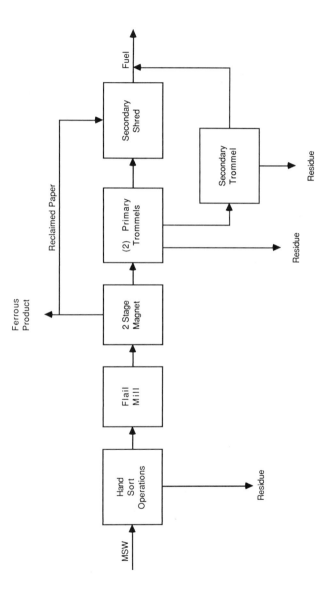

FIGURE 2. Moderate refuse derived fuel production flowsheet as employed by Combustion Engineering at Hartford, Connecticut.

-4 inches (-10.2 cm). Fuel produced by this system is then
burned in a CE spreader-stoker boiler.

The processing system employed by the Hartford plant
accomplishes a measurable improvement in fuel quality (Table
VII). The moderate RDF process employed at Hartford, Connecticut, does increase the quantity of waste being directed to
landfill, however.

b. *Gallatin, Tennessee.* A second moderate RDF process
has been installed at Gallatin, Tennessee, adjacent to the
Westinghouse-O'Connor waterwall rotary kiln incinerator. This
facility installed after the incinerator was built, uses an NRT
drum or AMSORT[R] process (see Figs. 3 and 4) to produce fuel,
ferrous metals, and reject. A secondary process recovers
aluminum as a salable product. The process itself is a simple
segmented drum into which raw refuse is fed directly. In the
first stage, cutters break the bags and expose the refuse to
processing. In the second stage, permanent magnets located
on the inside diameter of the drum remove ferrous metals from

TABLE VII. Ultimate Analysis of Raw Refuse and RDF
Projected at the Hartford, Connecticut, faility

Element/Compound (wt %)	Raw refuse	Prepared RDF (b)	(c)
Carbon	28.57	35.33	34.0
Hydrogen	2.91	3.60	4.1
Oxygen	17.26	21.60	23.47
Nitrogen	0.36	0.45	0.5
Sulfur	0.10	0.12	0.23
Chlorine	–[a]	–[a]	0.70
Moisture	28.80	25.90	23.0
Ash and inerts	21.80	13.00	14.0
Higher heating value			
Btu/lb	4600	5700	5785
Kcal/g	2.55	3.16	3.21

[a]Not reported. [b]Hazzard and Tease [9]. [c]Mirolli,
Ferguson, and Bump [18].

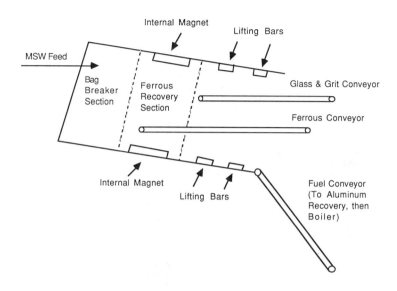

Figure 3.   Schematic of the NRT fuel production process.

the mass of waste. In the third and final stage, small lifters
within the drum remove glass and rocks from the refuse product.
Conveyors penetrating the drum recover the various product
streams. The final fuel product is shown in Table VIII.

The NRT system, like the trommel, has differential material
and fuel recovery rates depending upon type of material processed and operating rate (TPH). When the system is operating
at 7.7-12.3 (7.0-11.2 tonne/hr) the following recovery rates
have been experienced: Paper, plastics, textiles, etc. (as
fuel), 97.1-97.6%; wet organics such as food and yard waste
(as fuel), 31-41%; ferrous metals (as product), 78-90%; glass
(as reject product), 81-87%; and fines (as reject product),
49-85% [31, 32]. Unlike the trommel system, the NRT process
does not control fuel particle size.

The NRT system, like the CE trommel system, provides for
significant improvement in fuel quality. As shown in Table
VIII, the heat content of the Sumner County, Tennessee, waste
used as fuel improved by 32%, the moisture content decreased
by 9%, and the ash content decreased by 51% when the NRT segmented drum was installed. Table VIII also provides

FIGURE 4. Photo of NRT drum in operation. Note the conveyors extending into the drum. Note also the bag breaking teeth in the first portion of the drum.

TABLE VIII. Comparison of Raw Refuse and Fuel Produced by the NRT Process

Parameter	Weight percent	
	Raw refuse	NRT fuel
Volatile matter	41.2	54.5
Fixed carbon	4.4	5.9
Ash and inerts	23.1	11.3
Moisture	31.3	28.4
Sulfur	0.09	0.11
Higher heating value		
Btu/lb	4448	5865
Kcal/g	2.47	3.26

Source: [32].

fuel compositions based upon the tests performed at the Gallatin facility.

Moderate RDF processes provide for some control over heavy metal emissions. Data presented by Sommers et al. [31, 32] demonstrate that the NRT process may reduce the heavy metal content of MSW as follows: lead, 52%; cadmium, 73%; and chromium, 63%. These estimates are based upon reductions in uncontrolled airborne emissions [10, 31, 32].

2. *Conclusions Concerning Moderate RDF Production Processes*

The production of "moderate" grades of RDF improve the quality of the fuel as measured by calorific value, moisture content, ash content, and metals content, maximizing Btu recovery for the combustion process. The production of moderate type RDF products, however, introduces the element of a landfill burden from the fuel production. Inerts removed from the fuel stream are not recovered as product, but are sent directly to landfill.

C. Refined or "Fluff" RDF Production Processes

The final grade of fuel considered here is refined or
fluff RDF produced by trommel, by air classification, or by
screening processes. The method of analysis here uses case
studies and general analytical data.

1. *Case Studies in Refined RDF Production*

In order to evaluate this grade of fuel, two case studies
are employed: (1) Madison, Wisconsin; and (2) Kovik, Sweden.
The Madison plant employs a trommel based approach while the
Kovik plant is built around the Brini screen.

a. *Madison, Wisconsin.* The Madison plant was first to
apply a trommel to RDF production. It employs the same trommel design as CE uses at Hartford, Connecticut. The primary
difference between the Madison and Hartford plants is the fact
that there is only one trommel. Further, both primary trommel
underflow streams are reject product in Madison (Fig. 5).
Product recoveries at the Madison plant are as follows: glass
and dirt (as reject), 80%; ferrous metal (as product), 90%;
yard waste (as reject), 50%; fines (as reject), greater than
95% [29, 30]. Consequently, the mass recovery rate of RDF as
fuel is about 53-57% [29, 30, 37]. Energy recovery is about
65% [37].

As one of the earlier RDF plants, the Madison facility is
small: 400 TPD (360 tonne/d) of capacity. However it has
shown the ability to produce a useful fuel, as seen in Table
IX. Further, it demonstrates long term reliability, with
plant availability averaging over 93% since 1979 [29, 30].
Capacity factors are limited only by the availability of
refuse.

b. *Kovik, Sweden.* The Brini process developed by PLM
Selsberg, and installed both in Kovik and Sundsvall, Sweden,
offers a useful alternative to the trommel based systems discussed previously. In the Brini process installed at Kovik
(see Fig. 6), waste is coarsely shredded and directed to an
inclined, vibrating screen shown pictorially in Fig. 7. This
screen separates light paper and plastics, and other combustibles, into a fuel fraction. The light fraction migrates up
the screen to an outfeed conveyor (see Fig. 8). Large heavy
objects such as ferrous metals and heavy plastics flow downward to the base of the screen where they are discharged.
After magnetic separation, this fraction is landfilled. Heavy
small objects fall through the 3/4 inch (1.9 cm) openings in
the screen. This fraction is largely heavy organics such as

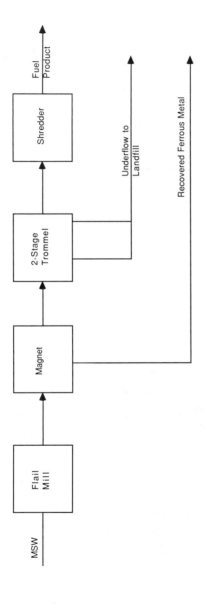

FIGURE 5. *Summary process flowsheet of the Madison, Wisconsin, process for producing "fluff" RDF.*

TABLE IX. Ultimate Analysis of Refined RDF Produced
by a Trommel Process (Basis: Madison, Wisconsin,
Flowsheet Applied to Reference Fuel)

Element/Compound	Weight percent
Carbon	32.2
Hydrogen	4.3
Oxygen	25.3
Nitrogen	0.37
Chlorine	0.58
Sulfur	0.79
Inerts	15.9
Moisture	21.2
Total	100.6
Higher heating value	
Btu/lb	5650
Kcal/g	3.14
Mass yield (%)	67.0
Energy yield (%)	78.0

food waste and yard waste, and glass. It is mixed with night
soil and sewage sludge, and composted. The process recovers
about 60% of the mass of waste as fuel, and captures about
80% of the incoming energy content of the MSW for use in the
fuel product. Another 20% of the mass of material is directed
to compost production, while the remainder is produced as
ferrous metal or landfilled [24].

Secondary processes have been installed at Kovik, also,
including fuel drying using available landfill gas as the
source of heat for the moisture process; and fuel briquetting
(see Fig. 6). Such processes are not employed at dedicated
RDF production plants collocated with combustion systems.
Consequently, Sundsvall plant burns the fluff RDF which is
neither dried nor pelletized [24].

The fuel product produced by the Brini process represents
a significant improvement when compared to raw refuse, as

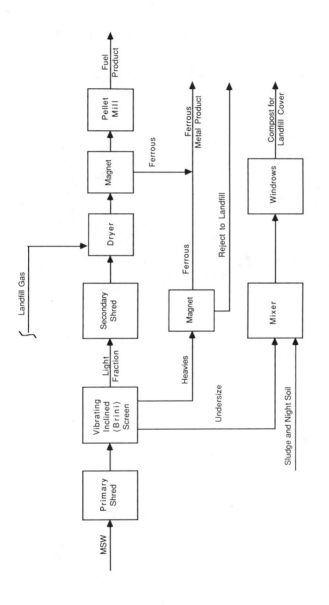

FIGURE 6. Summary process flowsheet of the Kovik, Sweden, RDF plant.

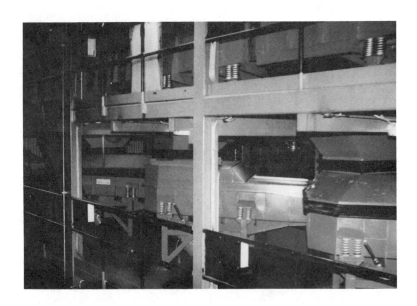

*FIGURE 7.   Photo of the bank of Brini screens at
Kovik, Sweden.*

shown in Table X [5, 21, 22].   Further, PLM Selsberg states
that the fuel contains only 10% inerts, and less than 1%
metals [24].   This product, like other "fluff" or refined RDF
products, provides for significant reduction in heavy metals
being sent to the combustor.

2.   Conclusions Concerning Refined RDF Production

Refined or fluff RDF offers a third option for producing
a combustible fuel from MSW.   It provides for the most extensive improvement in the quality of the fuel as measured by
heat content, moisture content, ash content, and heavy metals
content.   At the same time, however, the production of refined
RDF fuels does not maximize energy recovery from the waste
stream, and places an increasing burden on the landfill.

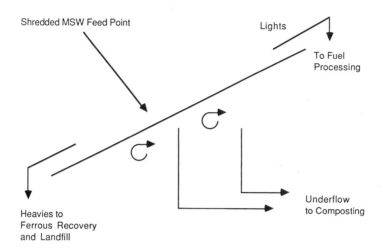

Shredded MSW Feed Point

Lights

To Fuel
Processing

Heavies to
Ferrous Recovery
and Landfill

Underflow
to Composting

FIGURE 8. Schematic of the Brini screen in operation.

D.   Overall Summary of RDF Production Technologies

     It is useful to compare the alternative RDF production
techniques available in terms of fuel and energy recovery,
and fuel quality, as well as the creation of landfill burden.
Approximate recovery rates, fuel ultimate analyses, and landfill burdens for various generic types of RDF production are
compared in Table XI. This table dramatizes the continuum
associated with RDF production technologies. In reality, one
can create processes at any intermediate point on the continuum by adjusting the configuration of the specific processes.
     It is also useful to consider the impact of RDF production
technologies on the environmental consequences of fuel combustion. It is readily apparent from the case studies that pro-
gression along the continuum towards refined RDF will minimize
the heavy metals being directed to the combustor. RDF, then,
is a viable alternative to the direct use of the mass of municipal waste as a feedstock for incineration. Further, RDF is
beginning to be viewed as a marriage between incineration and
recycling, particularly when hand picking is incorporated
into the fuel processing. Because RDF processes dramatically

TABLE X.   Comparison of Brini Refined RDF to Household
Waste

| | Weight percent | |
Component	Household waste	Brini RDF
Paper	36	70
Plastics	8	10
Food and yard waste	33	–
Metals	4	–
Glass	8	–
Wood	1	–
Textiles	5	16
Miscellaneous	5	4
Moisture	33	15
Ash	21	10
Higher heating value		
Btu/lb	4730	7310
Kcal/g	2.63	4.06

Source: Bergstrom [5].

alter the composition and nature of the fuel being burned,
the combustion technologies selected can be distinctly different from the machines described in Chapter III.

III.   THE COMBUSTION OF REFUSE DERIVED FUELS

Refuse derived fuels can be burned on the grates or in the
rotary kilns of systems designed for the mass burning of municipal waste, as is done in Kovik, Sweden; and Gallatin,
Tennessee.   Production of a more homogeneous fuel, with improved compositional characteristics, provides alternative
combustion system options, that can improve the efficiency of
energy recovery from MSW.   Typical RDF combustion options include spreader-stoker and fluidized bed boilers.   Both

TABLE XI.   Summary of Fuel and Product Recoveries,
    Ultimate Analyses, and Landfill Burdens as a Function
    of RDF Process (Basis: 1 Ton incoming refuse, based on
    calculations from reference MSW)

Analytical parameter	Process			
	Mass burn	Coarse	Moderate	Refined
Incoming waste (lbs)	2000	2000	2000	2000
Outgoing fuel (lbs)	2000	1900	1600	1340
Fuel yield (mass %)	100	95	80	67
Ultimate Analysis (wt %)				
Carbon	27.5	28.4	30.7	32.2
Hydrogen	3.7	3.8	4.1	4.3
Oxygen	20.6	21.8	23.7	25.3
Nitrogen	0.45	0.46	0.41	0.37
Chlorine	0.50	0.51	0.57	0.58
Sulfur	0.83	0.86	0.84	0.79
Inerts	23.4	20.3	18.4	15.9
Moisture	23.2	24.4	21.9	21.2
Total	100.2	100.5	100.6	100.6
Higher heating value				
Btu/lb	4830	5000	5410	5650
Kcal/g	2.68	2.78	3.00	3.14
Fuel yield (energy %)	100.0	98.0	90.0	78.0
Ferrous product yield				
Pounds	0 [a]	96	96	96
Percent of incoming MSW	0.0	4.8	4.8	4.8
Landfill burden				
Pounds	0 [b]	0	304	564
Percent of incoming MSW	0	0	15.2	28.2

[a] Typically ferrous metal is recovered from the bottom ash
of a mass burn incinerator.

[b] Does not include landfill burden from the combustion or
incineration process.

technologies are discussed below, focusing on case studies
fundamental information. While many RDF plants co-fire coal
or biomass and municipal waste based fuels (with coal or biomass being the supplement); supplementary suspension firing
of RDF in large coal-based utility boilers has become less
common in recent years.

A.   Spreader-Stoker Combustion of Refuse Derived Fuel

In a spreader-stoker (see Fig. 9), RDF is distributed
over the grate by means of a wind swept spout or air blown
stoker. The RDF is distributed as uniformly as possible.
Fine, dry particles burn in suspension above the grate.
Larger particles may ignite in suspension, but usually burn
out on the grate. Combustion gases pass from the furnace
area where combustion dominates, to the radiant and convection
sections of the boiler where heat recovery is the predominant
activity. Combustion gases, cooled by the superheater,
boiler, and economizer sections, then pass through the air
heater to the air quality control system.
Spreader-stoker combustion of RDF is practiced at the
following locations: Columbus, Ohio; Hooker Chemical, Niagara
Falls, New York; Albany, New York; Ames, Iowa; Akron, Ohio;
Haverhill, Massachusetts; Dade County, Florida; and Hartford,
Connecticut. These spreader-stoker boilers are highly similar
to the many wood fired units which have been built from
Burlington, Vermont, to Longview, Washington [36, 38, 39].
They represent technology transfer from the pulp and paper
industry to the waste processing industry.

1.   Case Studies in the Spreader-Stoker Firing of RDF

The coarse RDF spreader-stoker installation at Columbus,
Ohio, and the moderate RDF spreader-stoker installation at
Hartford, CT provide significant insights into this technology.
Consequently, they are described below.

   a.   Columbus, Ohio.   The Columbus plant was commissioned
in 1983 as a dedicated RDF process and boiler plant generating
electricity. The unit fires about 1,600 TPD (1,450 tonne/d)
of coarse RDF. This fuel is expected to provide at least
80% of the heat input to the boilers. The remainder of the
fuel is bituminous coal. RDF is fed through windswept spouts.
Coal is fed separately onto the grate by means of paddle-
wheel spreaders. There are three Babcock and Wilcox boilers
at Columbus, each with Detroit Stoker grates and RDF spreaders.
Each boiler generates 165,000 lb/hr (75 tonne/h) of 700 psig/
725°F (47 atm/385°C) steam. Each furnace has a grate area of

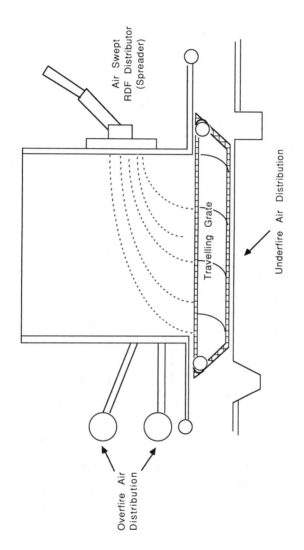

FIGURE 9. Schematic of the spreader-stoker firing system for RDF combustion. (Adapted from Seeker, Lanier, and Heap [27].

about 310 sq ft (29 m^2) and a volume of 16,517 cu ft (480 m^3).
Each boiler has a firing rate of 280 - 300 × 10^6 Btu/hr (265 -
285 GJ/h), the units have grate heat release rates of about
900 × 10^3 Btu/sq ft-h (10.26 GJ/m^2-h), and volumetric heat
release rates of about 17.5 × 10^3 Btu/sq ft-h (651 MJ/m^3-h)
of furnace [20]. The units are operated at a target excess
air level of 60%, and a target air distribution of 60% under-
fire/40% overfire air. Air control is not precise, however,
and 80% excess air is commonly experienced [20]. Combustion
air is preheated slightly. The Columbus facility has an
availability of 85-90%.

The Columbus facility was tested for NO$_x$ and particulate
emissions in February 1984 [24]. This test demonstrated the
effectiveness of the boiler and 4-field electrostatic precipitator (ESP), sufficient for regulations under which the plant
was constructed, when fired with 80% RDF, 20% coal. The
results showed a particulate emission rate of 103 lb/Btu
× 10^6 (12.9µg/GJ) after the ESP. Emissions of NO$_x$ were
shown to be 0.31 lb/Btu × 10^6 (13.3 µg/GJ) during that test
(Table XII).

   b. *Hartford, Connecticut.* The Hartford plant [9, 18, 29,
30] is designed to incinerate a moderate RDF as previously
described. The combustion systems are CE VU-40 boilers, each
capable of 231,000 lb/hr (105 tonne/h) when fired with RDF.
Any two boilers can carry the entire plant load. The steam
generated by these boilers is 880 psig/825°F (60 atm/440°C).
Typical unit firing rates on RDF are 56,300 lb/hr (25.5 tonne/h
of 5,785 Btu/lb (13.5 GJ/tonne) fuel, or 326 × 10^6 Btu/hr
(340 GJ/h). Each boiler burns about 675 TPD (610 tonne/d) of
RDF, which makes it equivalent to an 850 TPD (770 tonne/d)
mass burn unit when the materials recovery associated with RDF
production is considered.

The Hartford plant is operated with excess air levels on
the order of 40-60%. The unit is capable of firing 100% RDF,
without supplementary fuel, in its boilers. The thermal efficiency of this combustion system is expected to be on the
order of 75% [9, 18]. The Hartford plant is the most recent
RDF combustion facility using spreader-stoker technology to
be put on line. Consequently, it includes advances in the
technology, particularly in the post-combustion control area.
The unit uses a spray dryer-absorber and baghouse combination
for acid gas and particulate control (Chapter VIII).

2. *Manipulating Combustion of RDF in a Spreader-Stoker*

   Given the experience with spreader-stoker combustion of
RDF, considerable studies have been performed concerning

TABLE XII. Summary of Columbus, Ohio, Source Test for
Particulates and Oxides of Nitrogen

Firing Conditions

Coal - 20% of heat load

RDF - 80% of heat load

Load - 166,000 lb/h steam (single boiler)

Fuel quality (average of 3 runs)

	Moisture (%)	Ash (%)	Higher heating value Btu/lb	Kcal/g
Coal	5.2	7.22	12,477	6.92
RDF	25.3	21.0	5,597	3.11

Airborne emissions

Particulate - 7.41 lb/h (after ESP) filterable
- 152 lb/h (after ESP) condensable

- 0.03 lb/Btu $\times$ 10^6 filterable
- 0.615 lb/ Btu $\times$ 10^6 condensable

Oxides of nitrogen - 0.31 lb/Btu $\times$ 10^6

Source: Scheffel [25].

manipulation of the combustion mechanism. Particular attention has been devoted to the issues of combustion air, rates
of heat release, and airborne emission control.

a. *Combustion Air.* Because the level of excess air, the
distribution of air, and the temperature of air impact boiler
efficiency, combustion temperatures, and airborne emission
formation; these issues have been studied extensively by numerous authors. With respect to airborne emissions, particular
attention must be given to solids and inerts exiting the
boiler as flyash rather than bottom ash, emissions of heavy
metals, and formation and destruction of dioxins and furans.
Numerous authors have recommended various levels of excess
air for RDF-fired spreader-stokers. Authors such as
Massoudi [17], and Hazzard and Tease [9] have recommended
40-60% excess air (or 6.0-7.9% O_2 in the dry stack gas) as an
optimal range. Because excess air governs flue gas volume,
and consequently duct sizing; minimization is important.

At the same time, however, sufficient air is required for completeness of combustion and airborne emissions minimization.
Consequently levels on the order of 50-60% are most commonly
seen.

Excess air and fuel composition are the dominant factors
in determining combustion temperature. Combustion temperature
is a determinant of efficiency, influences corrosion control
[9], and governs dioxin and furan formation and destruction.
For dioxin and furan minimization, firebox temperatures of at
least 1800°F (980°C) for at least 1-2 sec. are strongly
recommended [27]. Excess air, along with adiabatic flame
temperature and furnace configuration, lead to time/temperature
profiling of the furnace. When coupled with heat transfer surface availability and heat transfer coefficients, the combination can lead to calculated unit temperature profiles such as
the following: grate, 2700°F (1480°C), furnace section below
bustle, 2300°F (1260°C); furnace section exit, 1700°F (930°C);
and boiler exit (to economizer section), 650°F (340°C). Typical gaseous residence times include 0 sec at the grate, 0.7-
0.9 sec at the bustle, and 2.9-3.1 sec to the furnace exit.

Distribution of underfire and overfire air is as important
as level of excess air. Distribution of air facilitates CO,
HC, dioxins and furans, and NO_x control. Grate stoichiometric
ratios (GSR), as discussed in Chapter III, are one method for
expressing air distribution or underfire/overfire air ratios.
When units are run with GSR <1.0, they are using staged combustion. As shown in Chapter III, however, dioxin emissions
are minimized when the GSR >1.0. Since dioxin and furan
emissions tend to be higher with RDF systems than mass burn
systems (see Table XIII), this becomes important to combustion
control. Typical recommendations are for 60-70% undergrate
air, and 30-40% overfire air. For RDF systems, GSR can be
approximated by the following expression:

$$GSR = 1.0454 \times UFA/OFA - 1.4267 \times \phi \qquad (4-1)$$

For units operated at 60% excess air, 35-40% overfire air will
provide a GSR >1.0 while maximizing combustion control and
minimizing organic emissions. Further, since combustion temperatures of RDF systems rarely exceed 2700°F (1480°C), thermal
NO_x emissions are minimized and fuel NO_x emissions are more of
a problem. Air preheating is the final issue. Most designs
call for a minimum of air preheat. It is rarely used to
levels above about 350°F (176°C) in RDF fired boilers. All
of the combustion air and related issues can be summarized in
terms of combustion control (Table XIV).

TABLE XIII. Summary of Dioxin and Furan Emission Ranges
from RDF Facilities Compared to Mass Burn Facilities

	Emission concentration range $(ng/Nm^3$ at 12% CO_2)	
Pollutant	RDF Fired	Mass burn
2,3,7,8 TCDD[a]	0.522–14.6	0.019–62.5
2,3,7,8 TCDF[b]	2.69[c]	0.168–448
TCDD[d]	3.47–258	0.195–1,160
TCDF[e]	31.7–1,680	0.322–4,560
PCDD[f]	64.4–4,300	1.13–10,700
PCDF[g]	96.2–8,100	0.432–14,800

[a]2,3,7,8 Tetrachlorodibenzo-p-dioxin.

[b]2,3,7,8 Tetrachlorodibenzofuran.

[c]Data available for only one test.

[d]Tetrachlorodibenzo-p-dioxin.

[e]Tetrachlorodibenzofuran.

[f]Polychloronated dibenzo-p-dioxin.

[g]Polychloronated dibenzofuran.

Source: Seeker et al. [27].

  b. *Heat Release.* Relatively early RDF spreader-stokers
designs called for grate heat release rates approaching
1×10^6 Btu/sq ft-h (11.4 GJ/M^2-h) and furnace volumetric
heat release rates approaching 20×10^3 Btu/cu ft-h (745 MJ/
m^3-h). More recently these design parameters have been
changed to grate heat release rates on the order of 700–
800×10^3 Btu/sq ft-h (7.95–9.08 GJ/M^2-h) and volumetric heat
release rates on the order of 13×10^3 Btu/cu ft-h (484 MJ/M^3-h.
These grate heat release rates are considerably higher than
comparable values for mass burn units. Consequently, the RDF
fired boiler may be somewhat narrower and taller than the
mass burn boiler.

  c. *Airborne Emission Control.* Organic emissions such as
CO, HCs, and dioxins and furans can be minimized by proper

TABLE XIV. Good Combustion Practices for Minimizing
Trace Organic Emissions from RDF Combustors

Element	Component	Recommendations
Design	Temperature at fully mixed height	1800°F (982°C) at fully mixed height
	Underfire air control	As required to provide uniform bed burning stoichiometry (see text)
	Overfire air capacity (not necessary operation)	40% of total air
	Overfire air injector design	That required for penetration and coverage of furnace cross-section.
	Auxiliary fuel capacity	That required to meet start-up temperature and 1800°F (982°C) criteria under part-load operations
Operation/ control	Excess air	3-9% oxygen in flue gas (dry basis)
	Turndown restrictions	80-110% of design--lower limit may be extended with verification tests
	Start-up procedures	On auxiliary fuel to design temperature
	Use of auxiliary fuel	On prolonged high CO or low furnace temperature
Verification	Oxygen in flue gas	3-9% dry basis
	CO in flue gas	50 ppm on 4-hr average-- corrected to 12% CO_2
	Furnace temperature	Minimum of 1800°F (982°C) (mean) at fully mixed height
	Adequate air distribution	Verification tests

Source: Seeker et al. [27].

combustion through use and distribution of excess air.
Similarly, NO_x can be controlled by fuel production techniques
and by combustion air controls. Acid gases such as SO_2 and
HCl are formed strictly as a function of fuel composition.
The remaining significant issues, then, are particulates and
heavy metals emissions.

Particulate emissions are a function of the mass of inert
material entering the boiler, the particle size distribution
of fuel entering the boiler, the completeness of combustion,
and the gaseous velocities in the furnace and boiler section.
These parameters, when combined, determine the percentage of
solids exiting the unit as bottom ash vs. the percentage of
solids exiting the unit as flyash. RDF boilers generally
burn fuels with less inert material. However, the inert
material tends to be smaller in particle size. Further, the
furnace geometry and flame temperatures contribute to relatively high gaseous velocities through the furnace and boiler.
Consequently, RDF spreader-stokers may discharge as much as
67-80% of the solid products of combustion as flyash [4, 20,
25].

Heavy metals emissions are generally less in RDF combustion
than in mass burning operations. However the issue is made
somewhat more complex by the higher temperatures associated
with RDF fired units, and the smaller particle sizes of the
flyash from RDF fired units. These latter two variables result in flyash which can be significantly enriched by heavy
metals when spreader-stoker firing of RDF is employed. Actual
values will depend upon specific fuels and operating
parameters.

   d. *Conclusions Regarding Spreader-Stoker Firing of RDF.*
Spreader-stokers offer a distinct alternative to the mass
burning of MSW. They provide for units with closer air control, higher rates of heat release, and alternative opportunities for airborne emissions control. Such units trade off the
fuel processing and associated fuel losses and associated
landfill consequences with the improved combustor efficiency.

B. Fluidized Bed Combustion of Refuse Derived Fuel

   While spreader-stoker units are still the norm, fluidized
bed combustion of RDF is emerging as an attractive alternative.
There are numerous fluidized bed combustor types including:
(1) bubbling beds, (2) multi-bed combustors, (3) circulating
fluidized bed combustors, and (4) pressurized fluidized bed
combustors [21, 22]. The original fluidized bed application
to RDF was in Duluth, Minnesota, involving co-combustion of
RDF and sewage sludge [40]. More recently, fluidized bed

combustion of RDF has been installed in Sande, Norway, Eksjo,
Landskrona, Lidkoping, and Sundsvall, Sweden, and it has been
provided for in Eskilstuna, Sweden. It has been installed at
the French Island, Red Wing, and Wilmarth plants of Northern
States Power Company [28] and the Tacoma Light Division Steam
Plant No. 2, Tacoma, Washington. It has been proposed for
Erie, Pennsylvania.

Fluidized bed reactors are well described in the literature (see, for example, [13, 14, 15, 16, 26]). Bubbling
fluidized bed BFB combustors such as those installed at
Duluth, Tacoma, and Eksjo, all conform to the following basic
principles: the combustor consists of a reactor vessel, a
windbox, a feed system, a gaseous outlet system, and a solids
withdrawal system; and the primary reactor is further divided
into the bed area and the freeboard space (see Fig. 10). Bed
media is introduced into the reactor vessel and heated to
combustion temperatures. Air is directed to the reactor vessel
from underneath, typically at superficial velocities in the
bed of 3-10 ft/sec (0.9-3.1 m/sec). The air raises the bed
particles to such an extent that they behave as a fluid, and
appear to "boil" on the surface. The velocity of the air is
insufficient, however, to blow the bed particles out of the
reactor vessel. Fuel and potentially other reactants (i.e.,
limestone for acid gas control) are introduced either within
the bed or over the bed. The BFB accomplishes combustion in
an environment that promotes maximum mixing and turbulence
among the reactants. At the same time the bed media (i.e.,
sand) intimately contacts the fuel particles, wearing away
combusted surfaces and exposing fresh fuel to the oxidation
process.

Fluidized bed reactors gain significant advantages from
the turbulence within the reactor vessel, and from the improved
heat transfer associated with bed particles intimately con-
tacting fresh fuel. Further, fluidized beds provide for in-
situ control of acid gases and avoidance of post-combustion
controls by promoting the following reactions:

Limestone calcination: $CaCO_3 \Rightarrow CaO + CO_2$ (4-2)

SO_2 capture: $CaO + O + SO_2 \Rightarrow CaSO_4$ (4-3)

HCl capture: $CaO + 2HCl \Rightarrow CaCl_2 + H_2O$ (4-4)

Temperatures in fluidized bed reactors are moderated by the
extent of heat removal from the combustor. Heat removal may
be accomplished either by in-bed boiler tubes for highly
efficient heat transfer or extensive waterwalls along the
sides of the reactor. Temperatures are typically maintained

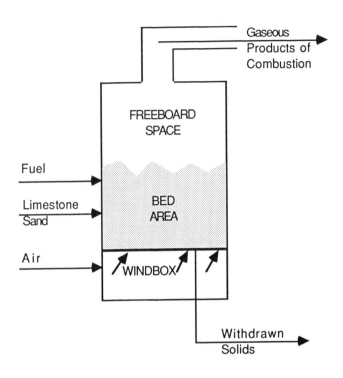

FIGURE 10. *Schematic of bubbling type fluidized bed.*

at about 1500°F (815°C) for optimal efficiency of the acid gas control reactions, and for elimination of serious problems associated with formation of thermal NO_x.

Circulating fluidized beds (CFB) differ from conventional fluidized beds in that the superficial velocities are much higher. Consequently, the bed is deliberately "blown out" with bed superficial velocities in the 15-20 ft/sec (4.6-6.2 m/sec) region and some fast reactors in the 25-30 ft/sec (7.7-9.2 m/sec) region. All of the bed media, fuel, and sorbent materials pass from the reactor vessel into a hot cyclone where gaseous products of combustion are separated from solid particles. Solid particles, so captured, are returned to the reactor vessel for additional combustion activity (Figs. 11, 12). Like conventional fluidized bed combustors, circulating fluidized bed reactors operate typically at about 1500-1600°F

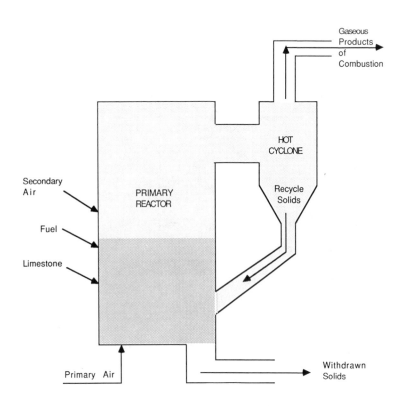

FIGURE 11. Schematic of circulating fluidized bed boiler.

(815-870°C) for maximum efficiency in acid gas control reactions. Recent CFB designs also have included staged combustion through complex combustion air distribution systems, and
ammonia injection, for NO_x control within the reactor.
    Fluidized bed technology offers a distinct alternative to
spreader-stoker technology in the combustion of RDF. Consequently, it is analyzed by use of both case study and more
fundamental data as a complete option for waste combustion.

1. Case Study: Sundsvall, Sweden

    The Sundsvall CFB facility includes a Brini RDF prepara-
tion plant making a refined, but not dried or pelleted, fuel
product. The boiler, built by Gotaverken, is relatively

FIGURE 12. Photo of hot cyclone at the Gotaverken CFB in Eskilstuna, Sweden; a unit designed for coal, wood, and RDF. The capture of solids in the hot cyclone, and their recycle to the combustor section of the reactor distinguishes the CFB from the bubbling bed.

small, having a capacity to generate 65,000 lb/hr (30 tonne/hr)
of 465 psig/520°F (30 atm/270°C) steam for use in cogeneration
application. The unit is designed to burn RDF as the dominant
fuel, however it also can fire wood waste, peat, hogged tires,
and other supplementary fuels. The Sundsvall CFB installation
is highly efficient, using waterwalls for heat removal, and
low levels of excess air (30-60% excess air, or 5-7% O_2 in the
wet stack gas).

Sundsvall is of interest not only for its efficiency, but
also its environmental consequence. Originally, the Sundsvall
plant used sand as its bed media, and only an ESP for air quality control. Recently, however, limestone has been used as a
sorbent at Sundsvall RDF [23]. Tests have been conducted on
the Sundsvall plant when operated with a wide variety of fuel
mixes. These tests, conducted prior to the use of limestone
as an acid gas sorbent are shown in Table XV. With use of
limestone, Gotaverken has demonstrated SO_2 and HCl capture
percentages of 95-98% when using limestone as a sorbent [7].
Consequently, Table XV also reports acid gases after 95%
reduction by the in-situ capture reactions discussed previously.
Additional tests for Sundsvall and a similar plant in Sande,
Norway, are shown in Table XVI [23]. The Sundsvall Plant generates very low levels of NO_x. Tests conducted have shown a
general trend conforming to the following equation (derived
from [12]):

$$NO_x \ (lb/Btu \times 10^6) = 0.055 + 0.004 \times EA(percent) \qquad (4\text{-}4)$$

Given excess air levels of 30-60%, the Sundsvall unit has
NO_x emissions of 0.175 - 0.295 lb NO_x/Btu × 10^6 (75 µg/GJ) of
heat input. Organic emissions, including dioxins and furans,
have been shown previously in Tables XV and XVI. CO emissions
ranged from 73 to 96 ppmv. Hydrocarbon emissions are generally
an order of magnitude lower than CO emissions [7]. Heavy
metals emissions have been reported as being an order of magnitude lower than those associated with mass burn units [24],
largely as a result of the fuel preparation process. The
Sundsvall plant and its sister facilities, including the
Sande, Norway, plant and the Eskilstuna, Sweden, plant have
demonstrated efficient combustion of RDF in a CFB. Fluidized
bed combustion is a realistic alternative to spreader-stoker
incineration of RDF.

2. Manipulating RDF Combustion in Fluidized Bed Reactors

As with spreader-stokers and mass burn units, the dominant
issues are combustion air control, heat release, and airborne
emissions control. These issues are considered below.

TABLE XV. Representative Airborne Emissions at the Sundsvall, Sweden, Plant (Prior to Use of Limestone for Acid Gas Control) (Values in mg/nm^3 except dioxins where values are in ng 2,3,7,8 TCDD eq./nm^3)

	RDF/ peat	RDF/ peat	RDF/ peat	RDF/chips tires
Fuel mixture (%)				
RDF	30-60	30-60	60-90	15-70
Peat	70-40	70-40	70-40	-
Wood	-	-	-	20-60
Tires	-	-	-	10-25
Emissions				
SO_x	251	177	197	226
(SO_x)[a]	12.6	8.9	9.9	11.3
HCl	122	188	489	281
(HCl)(a)	6.1	9.4	24.5	14.1
NO_x (as NO_2)	217	191	203	115
CO	35	39	251	24
HF	0.5	0.6	0.8	-
Dioxins	0.4	1.3	4.5	1.8
Combustion temp. (°F)	1562	1508	1526	1598
(°C)	850	820	830	870

[a]Calculated by authors at 95% in-bed capture of acid gases.
Source: [7]).

a. *Combustion Air Control.* Fluidized beds use the rate of heat removal, rather than excess air, to govern temperature. Further, they utilize mixing and turbulence through the action of the bed media to accomplish combustion and to eliminate the potential for cold spots and fuel rich pockets. Consequently, fluidized bed installations can use less excess air than spreader-stokers and mass burn installations. The Gotaverken unit in Sundsvall uses 30-60% excess air as

TABLE XVI.   Airborne emissions from the Gotaverken
Circulating Fluidized Bed Boiler at Sande, Norway

Parameter	Test results		
	1	2	3
Load (%)	80	80	80
Percent of fuel as RDF	100	100	100
Combustion temp. (°F)	1450	1500	1510
O_2 in wet stack gas (%)	5.0	5.0	4.7
CO (ppmv)	89	96	73
NO_x (ppmv)	129	110	168
SO_2 (ppmv)	60	67	87
Dioxins (TCDD eq. corr. to 10% CO_2)	0.7	1.4	0.5

Source: Oscarsson [23].

previously discussed.   The Red Wing and Wilmarth boilers of
Northern States Power Co., which are bubbling beds, use 50%
excess air [28].   Other values reported are within this same
range.   While some spreader-stoker units may approach 50%
excess air, the ability to operate in the 30% excess region
is unique to fluidized bed combustion systems.

   b.   Rates of Heat Release.   Rates of heat release for
fluidized bed units are comparable to or slightly higher than
those for spreader-stokers.   The Red Wing and Wilmarth units
are designed for heat release rates of 539×10^3 Btu/sq ft-h
grate equivalent (6.12 GJ/m^2-h) [28].   The Sande, Norway,
facility has a volumetric heat release of 28×10^3 Btu/cu ft-h
(1.04 GJ/m^3-h) [23].

   c.   Pollution Control Aspects.   The Sundsvall installation
demonstrates the ability of fluidized bed installations to
accomplish pollution control, particularly with respect to
organic emissions.   While Sundsvall uses an ESP, many recent
fluidized bed installations utilize fabric filters or baghouses for particulate control (see Chapter VIII).   Fluidized
bed airborne emission control is perhaps best illustrated by

the BACT determination filed for the Erie, Pennsylvania, plant (Table XVII). Note that the fluidized bed concept provides for control not only of organic emissions, acid gases, and oxides of nitrogen; but it also permits significant control of heavy metal emissions.

The fluidized bed combustion process, then, provides an emerging alternative to the spreader-stoker for the combustion of RDF. They are efficient, effective RDF combustors. Fluidized bed combustion also provides for significant combustion

TABLE XVII. Potential Airborne Emissions from Circulating Fluidized Bed Installations

	Emission rate	
	lb/ton RDF as fired[a]	lb/Btu $\times$ 10^6 [b]
Sulfur dioxide	2.4	0.2
Nitrogen oxides	3.0	0.25
Carbon monoxide	1.1	0.09
VOC (as methane)	0.06	5×10^{-3}
Lead	6×10^{-3}	5×10^{-4}
Beryllium (Be)	1×10^{-5}	8.3×10^{-7}
Mercury (Hg)	6×10^{-3}	5×10^{-4}
Fluorides (as HF)	3×10^{-3}	2.5×10^{-4}
Hydrogen chloride	1.1	.092
Arsenic (As)	1×10^{-4}	8.3×10^{-6}
Cadmium	1×10^{-3}	8.3×10^{-5}
Hexavalent chromium (as Cr)	5×10^{-4}	4.2×10^{-5}
Nickel	2.5×10^{-4}	2.1×10^{-5}
2,3,7,8 TCDD eq.[c]	4×10^{-8}	3.3×10^{-9}

[a]Conversion factor for kg/tonne RDF is 2.43.

[b]Assumes 12×10^6 Btu/ton of RDF; conversion to kg/GJ is 2.09.

[c]EPA method.   Source: [7].

air control and, if CFB combustion is employed, for staged
combustion. Further, fluidized bed systems provide for high
rates of heat release, and for substantial control of airborne emissions without extensive use of post-combustion technology.

IV. COSTS ASSOCIATED WITH REFUSE DERIVED FUEL
     TECHNOLOGIES

Numerous studies have been conducted concerning the costs
of RDF systems. In general these studies have shown that RDF
systems are comparable in cost to mass burn systems. Presented below is a summary of capital and operating cost data
associated with RDF plants.

A. Capital Costs

Detailed studies on capital costs have been performed for
the production facilities associated with processing the waste
into fuel, and for facilities which both produce and burn RDF.
The most detailed studies have been conducted by Kerstetter
and Schlorff [11]. They show that RDF production facilities
generally cost between $14,200 and $51,000 per daily ton of
MSW processing capacity, with a median of $23,600/daily ton of
processing capacity (1985 dollars). That range, updated to
1988 dollars is $15,900-$57,400/daily ton of capacity with a
median value of $26,600/daily ton of capacity.
Complete facilities including both the fuel processing and
the energy recovery facilities, with energy being produced as
electricity, have a cost range essentially identical to that
of a mass burn plant. The 1988 dollar cost median derived
from Kerstetter and Schlorff [11] is $79,700/daily ton of
capacity.
Our independent studies demonstrate identical conclusions
regarding RDF plant capital cost. An analysis of plants proposed, with comparable air quality control systems, yielded
the following capital cost equation:

$$K_{rdf} = 11.42 + .098 \times C \ (r^2 = 0.93) \qquad (4-5)$$

Where K_{rdf} is the capital cost of the facility in millions of
1988 dollars, and C is the capacity in TPD. The equation was
based upon nine proposed plants, including both spreader-
stoker and fluidized bed designs. This equation is highly
similar to the capital cost equation for mass burn facilities,

reported in Chapter III. When the two data sets are combined, they yield a cost estimating equation as follows:

$$K_{wte} = 14.77 + .094 \times C \ (r^2 = 0.88) \qquad (4-6)$$

The capital cost equations demonstrate that the final capital cost of a facility is less a function of technology than a function of the risks perceived on the part of investors.

B. Operating Costs

Operating costs for RDF facilities include labor for both fuel production and fuel consumption. Fuel production alone has an operating and maintenance cost of about \$20-\$28/ton in 1988 dollars [11]. When both fuel production and combustion are included the costs are on the order of \$25.7-\$38.5/ton processed (1985 dollars), or \$29-\$43/ton processed in 1988 dollars. The operating and maintenance costs associated with RDF systems are higher than those associated with mass burn systems. This differential declines substantially in larger systems (i.e., greater than 1,500 TPD).

V. CONCLUSIONS

Refuse derived fuel technologies offer a distinct, commercially proven alternative to mass burn technologies. Fuel can be produced at grades ranging from minimal processing (coarse RDF) to highly refined (fluff RDF). These grades involve heat content, moisture and ash content, and heavy metals content. RDF combustion technologies offer tighter combustion control due to the potential for using a more homogeneous fuel. As such, both spreader-stokers and fluidized beds are realistic, potentially cost-effective options.

REFERENCES

1. Anderson, L. L., and Tillman, D. A. (eds.). 1977. Fuels From Waste. Academic Press, New York.

2. Anon. 1988. Special Section: Energy From Wastes. Power 132(3): W-1 - W-24.

3. Barton, J. 1982. Evaluation of Trommels for Waste to Energy Plants, Phase I. Report of the Doncaster and Byker Test Series. National Center for Resource Recovery Washington, D.C. for U.S. Dept. of Energy. Contract No. DE-AC03-80cs24315.

4. Bechtel. 1975. Fuels from Municipal Refuse for Utilities: Technology Assessment. San Francisco. Prepared for Electric Power Research Institute, Palo Alto, California.

5. Bergstrom, R. 1987. Energy Recovery from Waste and Industrial Refuse. Stockholm Energi.

6. Ebasco. 1982. Technology Assessment: Municipal Solid Waste as a Utility Fuel. Electric Power Research Institute, Palo Alto, California.

7. Erie Energy Recovery Company. 1986. Technical Filing in Support of Best Available Control Technology (BACT). Determination for Erie Municipal Waste-to-Energy Plant.

8. Hasselriis, F. 1983. Thermal Systems for Conversion of Municipal Solid Waste, Vol. 4: Burning Refuse-Derived Fuels in Boilers: A Technology Status Report. Argonne National Laboratory, Chicago.

9. Hazzard, N. D., and Tease, H. V. 1986. Factors in the Selection of Boilers for Firing Refuse Fuels. Presented at the Arizona Water and Pollution Control Association Conference, Phoenix, Arizona.

10. Kenny, G. R., Sommer, E. J., and Kearley, J. A. 1986. Operating Experience and Economics of a Simplified MSW Materials Recovery and Fuel Enhancement Facility. Proc: 1986 National Waste Processing Conference, American Society of Mechanical Engineers, Denver, Colorado.

11. Kerstetter, J., and Schlorff, E. 1987. Municipal Solid Waste to Energy: Analysis of a National Survey. Washington State Energy Office, Olympia, Washington.

12. Kit, R., O'Cilka, J. A., and Berry, J. R. 1986. Creative Energy Systems and Projects Using Renewable Fuel Sources. Proc: American Power Conference, Chicago, Illinois.

13. Makansi, J. 1986. Fluidized-Bed Boilers Live Up to Their Performance Goals. *Power 130*(3):59-60.

14. Makansi, J. 1987a. Special Report: Co-Combustion: Burning Biomass, Fossil Fuels Together Simplifies Waste Disposal, Cuts Fuel Cost. *Power 131*(7):11-18.

15. Makansi, J. 1987b. Special Report: Fluidized-Bed Boilers. *Power 131*(5):S-1 - S-16.

16. Makansi, J., and Schwieger, R. 1982. Fluidized Bed Boilers. *Power 126*(8):S-1 - S-16.

17. Massoudi, M. S. 1984. Feasibility of 100 Percent RDF
    Firing for Power Generation. Proc: 1984 National
    Waste Processing Conference, American Society of Mechani-
    cal Engineers.

18. Mirolli, M. D., Ferguson, W. B., and Bump, D. L. 1986.
    Mid-Connecticut Resource Recovery Project. Proc: Joint
    Power Generation Conference, Portland, Oregon.

19. Mitre Corp. 1983. Waste Composition Quantities and
    Analysis: San Diego, California.

20. Mullins, R. 1987. Personal communication--interview
    with D. Tillman during site visit, February 25, 1987.

21. National Energy Adminstration and the National Environ-
    mental Protection Board (of Sweden). 1986a. Energy From
    Waste: Summary of a Study by The National Energy Admin-
    istration and the Swedish Environmental Protection Board.
    Stockholm, Sweden.

22. National Energy Administration and the National Environ-
    mental Protection Board (of Sweden). 1986b. Energy from
    Waste Full Report. Chapter 10 as translated by
    Ms. H. Hasselriis. Stockholm, Sweden.

23. Oscarsson, B. 1987. Experience With a CFB Boiler at
    Sande Paper Mill, Norway. Proc: TAPPI Engineering Con-
    ference, September 24, 1986, pp. 499-507.

24. Peterson, S. 1987. Presentation to Swedish Trade Council
    Tour Group Concerning Brini Process. October 9.

25. Scheffel, D. 1984. Emission Test Report: City of
    Columbus Municipal Electric Power Plant Boiler No. 1.
    Columbus, Ohio. PEDCo Environmental, Inc. Cincinnati,
    Ohio.

26. Schwieger, R. 1985. Fluidized-Bed Boilers Achieve
    Commercial Status Worldwide. *Power 129*(2):S-1 - S-16.

27. Seeker, W. R., Lanier, W. S., and Heap, M. P. 1987.
    Combustion Control of MSW Incineration to Minimize
    Emissions of Trace Organics. EER Corporation, Irvine,
    California, for USEPA Contract No. 68-02-4247.

28. Skinner, M. F., and Bloomer, T. M. 1986. Red Wing and
    Wilmarth Steam Plants RDF Conversion. Joint Power
    Generation Conference. ASME/IEEE. Portland, Oregon.

29. Smith, J. L. 1986. Early and Current Systems Utilizing
    Refuse Derived Fuels. Combustion Engineering Co.,
    Windsor, Connecticut.

30. Smith, M. L. 1986. Operating Reliability of RDF Facili-
    ties. Proc: Conference on Burning Our Garbage; Issues
    and Alternatives. Golden West Section, Air Pollution
    Control Association, San Francisco, California.

31. Sommer, E. J., Kenny, G. R., Kearley, J. A., and Roos,
    C. E. 1985. Effects of MSW Preprocessing on Thermal
    Conversion of MSW in Mass Burn Incineration. Report of
    Phase I Research for U.S. Department of Energy,
    Contract No. DE-ACO5-84ER80177.

32. Sommer, E. J., Kenny, G. R., Kearley, J. A., and Roos,
    C. E. 1986. Mass Fired Energy Conversion Efficiency,
    Emissions, and Capacity. Proc: 1986 National Waste
    Processing Conference. American Society of Mechanical
    Engineers, Denver, Colorado.

33. Swedish Ministry of the Environment and Energy. 1987.
    Official Government Paper 1986/87:157 on Waste Manage-
    ment, etc. Stockholm, Sweden.

34. Swedish Trade Council. 1987. Combustion Technology:
    From Fuels to Flue Gas Cleaning. Stockholm, Sweden.

35. Tillman, D. A. 1977. Energy From Wastes: An Overview
    of Present Technologies and Programs. *In* Fuels From
    Waste (L. L. Anderson, and D. A. Tillman, eds.). Academic
    Press, New York, pp. 17-39.

36. Tillman, D. A. 1987a. Wood Energy. *In* The Encyclopedia
    of Physical Science and Technology, Vol. 14. Academic
    Press, San Diego, California, pp. 641-655.

37. Tillman, D. A. 1987b. Personal communications from site
    visits to Madison, Wisconsin; Columbus, Ohio; and
    Hartford, Connecticut refuse derived fuel projects.

38. Tillman, D. A. 1987c. Biomass Combustion. *In* Biomass:
    Regenerable Energy (D. O. Hall and R. P. Overend, eds.).
    John Wiley & Sons Ltd., London, pp. 203-219.

39. Tillman, D. A., Rossi, A. J., and Kitto, W. D. 1981.
    Wood Combustion: Principles, Processes and Economics.
    Academic Press, New York.

40. Trezek, G. 1983. Thermal Systems for Conversion of Muni-
    cipal Solid Waste, Vol. 6: Fluidized-Bed Combustion:
    A Technology Status Report. Argonne National Laboratory,
    Chicago, Illinois.

Chapter V

FUNDAMENTALS OF SOLID HAZARDOUS

WASTE COMBUSTION

I. INTRODUCTION

Incineration is a proven method for the permanent destruction of solid hazardous wastes. As such, it is a common element in managing hazardous wastes as demonstrated by the use of incineration in numerous permanent regional facilities. Examples of these facilities include the SCA installation in Chicago owned by Chemical Waste Management; the Rollins Environmental installations in Bridgeport, New Jersey, and Deer Park, Texas; SAKAB, AB, located in Nortorrp, Sweden; the HIM facility in Hessen, Germany; and Kommunekemi A/S, in Hyborg, Denmark.

Manufacturers such as Eastman Kodak, DuPont Chemical, and PPG also use incineration facilities for disposal of certain hazardous compounds. On-site incineration using transportable or mobile facilities is frequently considered as the method to treat solid hazardous wastes found on CERCLA[a] (or Superfund) sites. Although incineration may be expensive, it is considered to be the ultimate method for hazardous waste treatment and disposal. Typically, incineration is only one unit in a series of operations in a total hazardous waste management system (Fig. 1). Another potential installation is a mobile or transportable incineration system at a specific hazardous waste site (Fig. 2). Such uses of hazardous waste incineration are the consequent concerns of this text.

A. The Purpose of Solid Hazardous Waste Incineration

Hazardous wastes have been defined previously in Chapter I. As discussed in that chapter, combustible solid hazardous materials may include soils contaminated with solvents, petrochemical products, residues from coal gasification

[a]CERCLA is the abbreviation for the U.S. Government's program to clean up hazardous waste sites under the Comprehensive Environmental Response, Compensation, and Liability Act.

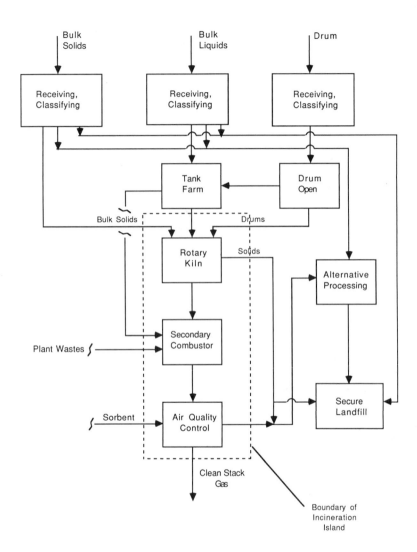

FIGURE 1. *Generalized flowsheet of a permanent hazardous waste incinerator.*

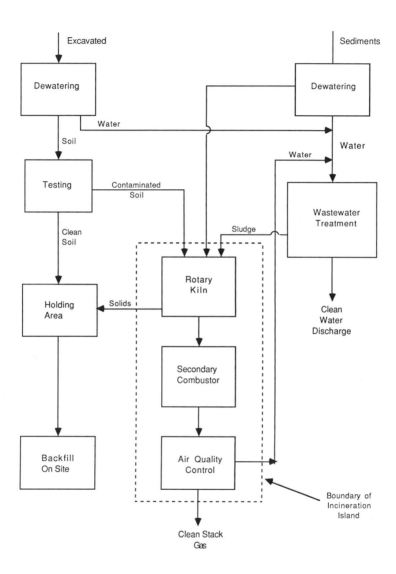

FIGURE 2. *Generalized flowsheet of a Superfund site (CERCLA) mobile or transportable incineration facility.*

(i.e., napthalene, anthracene), and a wide variety of other
organics. Solid hazardous wastes also include solid pesticide
materials such as aldrin, endrin, and atrazine; solid halocarbons such as hexachlorophenol; and numerous other compounds
and mixtures. Solid hazardous wastes can have sufficient
heating value to sustain combustion, or they can be contaminated dirts that must be heated and treated by use of supplementary fuels. In contaminated soils, the hazardous constituents may include not only organic compounds but also heavy
metals such as lead, zinc, copper, and vanadium. If landfilled, such hazardous wastes present a balance sheet liability for any company--essentially forever. The waste generator
is responsible for the wastes even at the landfill.

The purpose of incinerating solid hazardous wastes is quite
distinct from that of incinerating mixed MSW. Solid hazardous

Incineration presents a proven, permanent solution to such
hazardous waste disposal. Successful incineration, therefore,
reduces the liabilities of such hazardous wastes on the balance sheet. Further, for "orphaned" sites where the generator
is either no longer in existence or not found, incineration
reduces the risks associated with hazardous wastes from
society's balance sheet.

The purpose of incinerating solid hazardous wastes is quite
distinct from that of incinerating mixed MSW. Solid hazardous
waste incineration must meet legal and regulatory requirements
for essentially complete (99.99% - 99.9999%) destruction of
the contaminant in question, yet also achieve a significant
level of volume reduction. Incineration of MSW reduces volume
but also has the potential for cogeneration of electricity
along with the generation of additional revenue.

B.   Constraints on Hazardous Waste Incineration

The constraints on hazardous waste incineration are
legally imposed requirements associated with the degree of
destruction. The Resource Conservation and Recovery Act
(RCRA) states that incineration must destroy principal organic
hazardous constituents (POHCs) to a level of 99.99%. This
destruction and removal efficiency (DRE) requirement is measured as follows:

$$DRE = [(W_{in} - W_{stack\ gas})/W_{in}] \times 100 \qquad (5\text{-}1)$$

Where W_{in} is the mass flow of a given POHC into the incinerator, and $W_{stack\ gas}$ is the mass flow of the POHC measured in
the stack gas. At the same time the solids removed from the
incinerator must be subjected to EP Toxicity and/or TCLP testing, and potentially treated before being discharged to a
secure landfill. Hydrogen chloride (HCl) emissions from

incineration are required to be controlled to a level of <4 lb/hr, or collected at a scrubber efficiency of >99%, according to the RCRA statute [2,6].

The Toxic Substances Control Act (TSCA) governs incineration applied to polychlorinated biphenyl (PCB) oils and compounds. It mandates DREs of 99.9999% and imposes specific regime requirements on hazardous waste incinerators. Further, the regime requirements are explicitly spelled out by law (see Table I). While these legal limits are not legally imposed on incinerators used to dispose of hazardous wastes at Superfund[a] sites, the incineration installations (typically of the mobile facility variety) must meet the spirit of RCRA regulations.

Incineration systems also are required to minimize the formation and emission of products of incomplete combustion (PICs). Such products include hydrocarbons and halocarbons that are degradation products from the POHCs, or combinations of degradation products. PICs are commonly measured and considered in terms of carbon monoxide (CO) emissions, with CO serving as a surrogate or tracer compound.

State and local jurisdictions also impose legal constraints on incineration facilities. Such constraints usually focus on airborne emissions, liquid and solid discharges, and other environmental impacts. In the area of airborne emissions, the incinerator may be forced to comply with best

TABLE I.   Regime Requirements for Incineration of
           PCB Oils

	Condition No. 1	Condition No. 2
Gas residence time	2.0 sec.	1.5 sec.
Furnace (afterburner) temperature	2200°F (1204°C)	2950°F (1621°C)
Minimum O_2 in stack (dry basis)	3.0%	2.0%

Source: [6].

[a]Superfund is the acronym for the U.S. Government-funded program to dispose of hazardous wastes as mandated by the Comprehensive Environmental Response, Compensation, and Liability Act.

available control technology (BACT), as will be discussed in Chapter VIII. All of these constraints impact the design and operation of the incineration system.

C. Approach of This Chapter

Given the variety of solid hazardous wastes, the purposes of hazardous waste incineration, and the associated legal constraints, this chapter begins the technology evaluation process by first reviewing principles. It evaluates hazardous waste incineration through the application of combustion fundamentals.

First, solid hazardous wastes are treated as a collection of materials that can be thermally treated, whether or not the specific materials will sustain combustion. The chapter then examines the fuel properties associated with solid hazardous wastes including chemical structure, proximate and ultimate analysis (including presence of sulfur, nitrogen, chlorine, and other halogens, such as fluorine), calorific value, reactivity, and other thermodynamic properties such as heat capacity. Next, an examination of the combustion process is presented including a general mechanism, parameters impacting DRE (including the use of supplementary fuel), parameters impacting formation of PICs, and parameters impacting formation and control of airborne emissions such as acid gases, oxides of nitrogen, CO, and dioxins and furans. Because this chapter reviews much chemistry which is also presented in Chapter II, it provides abbreviated discussions where practical to minimize repetition.

II. COMBUSTION CHARACTERISTICS OF SOLID HAZARDOUS WASTES

As noted in Chapter I, a vast range of solid hazardous wastes are currently available and appropriate for incineration treatment. This range includes solid hazardous compounds plus soils contaminated with volatile and semi-volatile liquids and solids. A significant backlog of many sites in the U.S. and the industrialized world contain soil which has been contaminated with PCB oil. This backlog or inventory exists on CERCLA and Defense Environmental Restoration Act (DERA) sites, in temporary storage facilities, and in numerous other locations. Further, there are 290 million tons (264 million tonnes) of hazardous wastes generated annually in the U.S., of which 28 million tons (25 million tonnes) is amenable to incineration. The properties of such materials related to incineration are considered below.

A. Chemical Composition of Representative Solid
    Hazardous Wastes

    Chemical compounds considered as hazardous wastes include
a variety of solids, and volatiles and semi-volatiles within
a solid matrix (i.e., soils). Many of these compounds are
suitable for incineration.

1. *Structural and Compositional Considerations*

    Among the most common hazardous compounds suitable for
incineration are benzene, toluene, xylene, carbon tetrachloride, chloroform, trichloroethylene, perchlorethylene, chlorobenzene, pentachlorophenol, hexachlorobenzene, acetone,
methyl-ethyl-ketone (MEK), anthracene, and napthalene. Organic
pesticides which may be incinerated include aldrin, dieldrin
chlordane, DDD, DDT, picloram, malathion, atrazine, captan,
zineb, and mirex. Chemical compositions of several representative hazardous organic wastes are shown in Table II.
Representative structural formulae for select compounds are
shown in Fig. 3.
    Often hazardous compounds do not occur in isolation, or
in manufacturing settings. Rather, they occur as soil and
sediment contaminants. The Coleman-Evans site in Florida, for
example, is a CERCLA site with soils contaminated by pentachlorophenol. Numerous wood-treating facilities have such
contamination problems. The Bog Creek Farm site in Howell
Township, New Jersey, contains a mixture of contaminants from
paint manufacturing (Table III). Numerous other contaminated
sites contain mixtures of pesticides, dilute concentrations
of explosives (i.e., trinitrotoluene or TNT), byproducts of
coal gasification such as coal tars, PCB oils, and a host of
other products. In addition, many of the soils contaminated
with incinerable organics also contain measurable quantities
of heavy metals which must be accommodated in the incinerator.

2. *Ultimate Analysis and Higher Heating Value*
    of Hazardous Wastes

    Ultimate analyses and calorific values are common methods
for evaluating the fuel properties of combustible materials.
Calorific value is also used as the primary POHC ranking
method by the U.S. Environmental Protection Agency. Ultimate
analyses and higher heating values have been shown for municipal waste in Chapter II and are presented in Table IV for
selected hazardous wastes. Noteworthy is that the carbon,
hydrogen, sulfur, and chlorine contents of waste have independent influences on the heat of combustion or higher heating

TABLE II. Representative Hazardous Organic Chemicals

Compound name	Empirical formula	Molecular weight	State at 77°F (25°C) 1 atm
Aniline	C_6H_7N	93.12	Liquid
Benzene	C_6H_6	78.11	Liquid
Carbon tetrachloride	CCl_4	153.85	Liquid
1,2 Dichlorobenzene	$C_6H_4Cl_2$	147.01	Liquid
Hexachlorobenzene	C_6Cl_6	284.82	Solid
Nitrobenzene	$C_6H_5NO_2$	123.11	Liquid
Pyridine	C_5H_5N	79.10	Liquid
1,2,4-trichlorobenzene	$C_6H_3Cl_3$	181.46	Liquid
Atrazine	$C_8H_{14}N_5Cl$	215.69	Solid
Aldrin	$C_{12}H_8Cl_6$	364.94	Solid
Chlordane	$C_{10}H_6Cl_8$	409.83	Liquid
Dieldrin	$C_{12}H_8Cl_6O$	380.94	Solid
Dicyclopentadiene	$C_{10}H_{12}$	132.20	Solid
Hexachlorocyclopentadiene	C_5Cl_6	272.79	Liquid
Ethyl chloride	C_2H_5Cl	64.52	Gas
Methyl chloride	CH_3Cl	50.49	Gas
Methylene chloride	CH_2Cl_2	84.95	Liquid
Chloroform	$CHCl_3$	119.40	Liquid
Chlorobenzene	C_6H_5Cl	112.56	Liquid
Phenol	C_6H_5OH	94.11	Solid

Source: CRC Handbook of Chemistry and Physics.

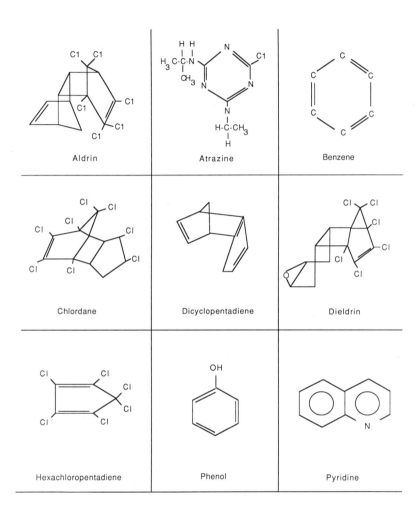

FIGURE 3. Structural formulae for selected hazardous waste compounds. Source: CRC Handbook of Chemistry and Physics.

TABLE III. Chemical Analytical Results: Waste Samples – Bog Creek Farm Site
Six Samples Collected by NUS Corporation October 3 – 4, 1984

PP No.	CAS No.	Compound	No. of occurrences	Concentration range (ppm)
Organics				
	67-64-1	Acetone	6	9 – 2,100
	78-93-3	2-butanone	3	16 – 5,200
44V	75-09-2	Methylene chloride	2	2 – 26,000
11V	71-55-6	1,1,1-trichloroethane	3	5,300 – 8,800
87V	79-01-6	Trichloroethene	3	4,700 – 5,500
85V	127-18-4	Tetrachloroethene	3	840 – 6,800
10V	107-06-2	1,2-dichloroethane	1	6,800
6V	56-23-5	Carbon tetrachloride	1	570
23V	67-66-3	Chloroform	1	550
4V	107-06-2	Benzene	4	30 – 8,900
86V	108-88-3	Toluene	6	8 – 180,000
	1330-20-7	Total zylenes	5	1 – 14,000
38V	100-41-4	Ethylbenzene	4	30 – 4,700
65A	108-95-2	Phenol	3	76 – 760
25B	95-50-1	1,2-dichlorobenzene	3	160 – 450

166

PP No	CAS No		n	Range (ppm)	
55B	91-20-3	Naphthalene	3	160 –	380
54B	78-59-1	Isophorone	4	39 –	890
61B	91-57-6	2-methylnaphthalene	2	21 –	88
	86-30-6	N-nitrosodiphenylamine	1		210
66B	117-87-7	Bis(2-ethylhexyl)phthalate	5	10 –	1,400
68B	84-74-2	Di-n-butylphthalate	4	82 –	1,400
67B	85-68-7	Butylbenzylphthalate	2	96 –	260
		Inorganics			
		Aluminum	5	80 –	2,610
		Barium	3	58 –	430
		Calcium	6	1,120 –	36,400
		Chromium	5	7 –	718
		Cobalt	3	6 –	27
		Copper	4	3 –	174
		Iron	5	876 –	5,160
		Lead	5	4.6 –	19,060
		Manganese	5	7 –	78
		Mercury	4	.27 –	2.2
		Senenium	1		6.4
		Thallium	1		(20)
		Vanadium	4	4 –	13
		Zinc	5	6 –	364

Notes: ppm   – (mg/kg).
        PP No. – Priority Pollutant Number.
        CAS No. – Chemical Abstracts Service Number.

TABLE IV. Ultimate Analyses and Higher Heating Values of Selected Hazardous Wastes

Compound name	Ultimate analysis					HHV (Btu/lb)	HHV (Kcal/g)
	C	H	Cl	O	N		
Aniline	0.7738	0.0758	–	–	0.1504	15,730	8.74
Benzene	0.9225	0.0075	–	–	–	18,046	10.03
Carbon tetrachloride	0.0781	–	0.9219	–	–	430	0.24
1,2-Dichlorbenzene	0.4902	0.0274	0.4824	–	–	8,222	4.57
Hexachlorobenzene	0.2530	–	0.7470	–	–	3,225	1.79
Nitrobenzene	0.5854	0.0409	–	0.2599	0.1138	9,896	5.50
Pyridine	0.7592	0.0637	–	–	0.1771	14,088	7.83
1,2,4-trichlorobenzene	0.3971	0.0167	0.5862	–	–	6,117	3.40
Atrazine	0.4455	0.0654	0.1644	–	0.3247	–	–
Aldrin	0.3950	0.0221	0.5830	–	–	6,750	3.75
Chlordane	0.2931	0.0148	0.6922	–	–	4,876	2.71
Dieldrin	0.3784	0.0212	0.5585	0.0420	–	10,004	5.56

Dicyclopentadiene	0.9085	0.0915	–	–	–	–
Hexachloro-cyclopentadiene	0.2201	0.7799	–	–	–	3,778	2.10
Ethylchloride	0.3723	0.0781	0.5496	–	–	–	–
Methylchloride	0.2379	0.0599	0.7022	–	–	–	–
Methylene Chloride	0.1414	0.0237	0.8349	–	–	–	–
Chloroform	0.1006	0.0084	0.8910	–	–	1,350	0.75
Chlorobenzene	0.6402	0.0448	0.3150	–	–	11,875	6.60
Phenol	0.7657	0.0643	–	0.1700	–	14,000	7.78

value of any compound. For nonhalogenated materials such
heats of combustion can be approximated by the Dulong formula
[11] as shown below:

$$HHV = 14,490 \ (c) + 61,000 \ (H) + 5,550 \ (S) \qquad (5-2)$$

Where C. H, and S are the weight fractions of carbon, hydrogen,
and sulfur in any given compound. The Dulong formula typically is within 3% of the actual heat of combustion for fossil
fuels. It may be less accurate for waste-based materials.
Its inaccuracy in evaluating some hazardous wastes results
from neglecting heats of formation for compounds, and from
its neglect of chemical structures--hence, the presence of
partially oxidized functionalities such as hydroxyls and
carboxyl structures, and halogenated structures.

The influence of chlorine, and the entire halogen family,
is somewhat more complex than the influence of carbon, hydrogen, and sulfur. Chlorine, or any other halogen, can act as
an oxidant for hydrogen. Alternatively, chlorine can be
evolved as Cl_2 gas, depending upon conditions in the incinerator. The influence of chlorine on the higher heating value of
selected organics is modest at relatively low concentrations,
but becomes severe at high proportions. At very high concentrations, chlorine prevents combustion (i.e., CCl_4').

In some cases a hazardous chemical spill can be absorbed
by a combustible absorbent such as corn cobs, a cellulosic
material. Sawdust and other biomass materials are also used
as absorbents. In such cases the mixture of chemicals and
absorbent will have a chemical composition, ultimate analysis,
and higher heating value proportional to the blend. The cellulosic material may have an empirical formula similar to cellulose: $C_6H_{10}O_5$. The higher heating value of cellulose is
7,500 Btu/lb. Alternatively, the absorbent may have an
empirical formula closer to that of dry wood, typically:
$C_{4.1-4.9}H_{5.1-7.7}O_{2.0-3-1}$ [24]. The higher heating value of
such material would be on the order of 8,000 - 9,000 Btu/lb
(dry). The total material resulting from absorbing a hazardous waste onto such a cellulosic based substrate will be
enriched in carbon and hydrogen, may contain a halogen, and
may have a higher calorific value than that reported for the
absorbent.

Ultimate analyses and higher heating values also are
measured for mixtures of materials such as soils and sediments
contaminated with chemicals and found at hazardous waste
sites. Representative values for the soils and sediments at
the Bog Creek Farm site (Table V) show that candidate materials
for incineration do not necessarily have to be capable of sustaining combustion. Rather, incineration may be considered

TABLE V. Ultimate Analysis of Soils and Sediments
at Bog Creek Farm Site

	Value	
Parameter	Soil (%)	Sediment (%)
Ultimate analysis		
Carbon (not including carbonates)	0.35	3.63
Hydrogen	0.13	0.10
Oxygen	2.66	2.71
Nitrogen	0.01	0.04
Sulfur	0.02	0.08
Moisture	13.03	51.11
Ash	83.80	42.84
Higher heating value (Btu/lb)	50.6	632.3
(Kcal/g)	0.03	0.35
Bulk density (lb/cu ft)	98.4	84.1
Heating capacity of inert dry solids (Btu/lb °F)	0.18	

simply as a means for soil and sediment thermal decontamina-
tion. Included in these variables must be the ash fusion
temperature of the inert portion of the solid hazardous waste,
including the soil matrix. Eutectic points and effects associated with the total mass of waste are essential values in
system selection and design. Finally, hazardous chemicals
may be absorbed onto porous inert materials with the consis-
tency of "kitty litter" as an alternative to cellulosic materials.

B. Reactivity of Selected Wastes

Reactivity is as important a combustion parameter as
ultimate analysis and higher heating value, particularly for
the incineration of hazardous wastes where the ability to
sustain combustion is not necessarily a criterion for select-
ing thermal destruction as the appropriate treatment. Reactivity, here, is used as a means for discussing the

temperature at which the complex thermal oxidation reaction
sequence commences, and the rate at which it may proceed.

For "pure" hazardous waste compounds such as pesticides,
tars (i.e., anthracene, napthalene), hexachlorobenzene, and
related compounds; thermogravimetric analysis (TGA) curves
are the best representations of reactivity. TGA curves for
aldrin and atrazine, two pesticide compounds of consequence
(Figs. 4 and 5) show that both compounds are highly reactive
and at very low temperatures (i.e., about 600°F). Further,
once decomposition reactions commence, they volatilize nearly
completely very rapidly.

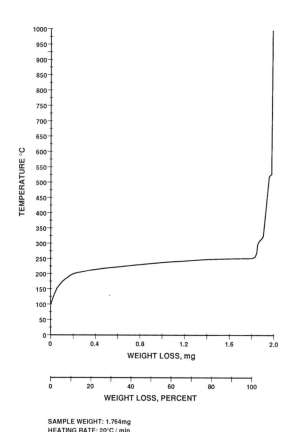

SAMPLE WEIGHT: 1.764mg
HEATING RATE: 20°C / min
ENVIRONMENT: ~ 20 ml / min air (20% O2, 80% N2)

FIGURE 4. Thermogravimetric analysis (TGA) curve for
recrystallized aldrin. (From Ferguson et al. [7].)

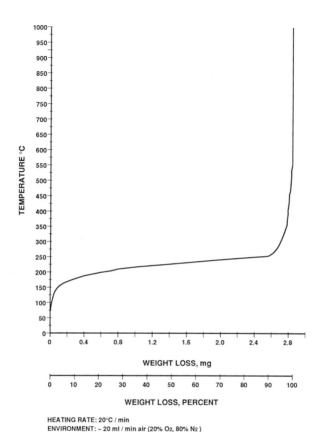

FIGURE 5. Thermogravimetric analysis (TGA) curve for technical atrazine. (From Ferguson et al. [7].)

For volatiles and semivolatiles mixed in a soil matrix, the data developed by Dellinger and coworkers [3,4] provides the most useful information. Dellinger and coworkers determined the temperature at which 20 hazardous compounds would begin to decompose (Table VI). The range of appropriate temperatures is from 730°F to 1400°F (390–760°C), with most compounds beginning the process of decomposition in the 1000°F (540°C) to 1200°F (650°C) range. These data also show that once reactions commence, they proceed vigorously to

TABLE VII. Reaction Initiation Temperatures for Various
Hazardous Wastes

Compound	Empirical formula	T_{onset}[a] °F	T_{onset}[a] °C
Acetonitrile	C_2H_3N	1400	760
Tetrachloroethylene	C_2Cl_4	1220	660
Acrylonitrile	C_3H_3N	1200	650
Methane	CH_4	1220	660
Hexachlorobenzene	C_6Cl_6	1200	650
1,2,3,4-tetrachlorobenzene	$C_6H_2Cl_4$	1220	660
Pyridine	C_5H_5N	1150	620
Dichloromethane	CH_2Cl_2	1200	650
Carbon tetrachloride	CCl_4	1110	600
Hexachlorobutadiene	C_4Cl_6	1150	620
1,2,4-trichlorobenzene	$C_6H_3Cl_3$	1180	640
1,2-dichlorobenzene	$C_6H_4Cl_2$	1170	630
Ethane	C_2H_6	930	500
Benzene	C_6H_6	1170	630
Aniline	C_6H_7N	1150	620
Monochlorobenzene	C_6H_5Cl	1000	540
Nitrobenzene	$C_6H_5NO_2$	1060	570
Hexachloroethane	C_2Cl_6	880	470
Chloroform	$CHCl_3$	770	410
1,1,1-trichloroethane	$C_2H_3Cl_3$	730	390

[a]Indicates temperature at which degradation reactions
commence.   Source: Dellinger et al [4].

complete decomposition. Data developed by Pershing and coworkers [13, 14, 29] have shown that soil particle size and
type impact these temperatures to some extent. Fine particle
soils such as clays require additional energy to initiate
reactions. Further, fine particles require substantial
additional energy to drive the volatiles completely

from the soil matrix. Fine particle soil incineration is
therefore more energy intensive.

C. Summary of Hazardous Waste Characteristics

    Solid hazardous wastes, then, come in a wide variety of
forms. They may or may not be capable of self-sustaining
combustion and may contain substantial quantities of such
heteroatoms as chlorine and other halogens, sulfur, and
nitrogen. A common link among solid hazardous wastes is significant reactivity at relatively low temperatures. This
reactivity is essential in the analysis of combustion mechanisms and systems.

III. THE PROCESS OF HAZARDOUS WASTE COMBUSTION

    Tsang and Schaub [27] have shown that thermal destruction
of hazardous waste can be simplified into the following
sequence of equations:

    unimolecular destruction

$$H_z => product(s) \qquad (5-3)$$

    bimolecular destruction

$$R + H_z => products(s) \qquad (5-4)$$

Where the rates of destruction are as follows:

    (for reaction 5-3)

$$dH_z/dt = -k_{uni}H_z \qquad (5-5)$$

    and (for reaction 5-4)

$$dH_z/dt = -k_b(R)(H_z) \qquad (5-6)$$

These lead to the following combined decomposition equation:

$$-dH_z/dt = k_{uni}[H_z] + (k_{bi}R_i)[H_z] \qquad (5-7)$$

Where k_{uni} is the unimolecular rate for decomposition and k_{bi}
is the bimolecular rate constant for attack of species R_i on
H_z. This sequence of equations provides an overall approach
to the kinetics associated with thermal decomposition of
hazardous waste.

In practice the incineration or thermal destruction of solid hazardous waste is a highly complex sequence of reactions that involves a myriad of significant steps. This process of hazardous waste incineration is reviewed below with particular emphasis on the following concerns: (a) mechanisms of hazardous destruction, (b) the influence of temperature on hazardous waste destruction, and (c) considerations associated with the formation/suppression of airborne emissions (including PICs) and solid residues.

A. Mechanisms of Solid Hazardous Waste Combustion

As developed by Dellinger and coworkers [3,4], and by other authors as well (i.e., [1, 21, 27, 28]), a global mechanism schematic can be developed for solid hazardous waste combustion which highlights key phases of solid hazardous waste incineration (see Fig. 6). These phases include vaporizing solids to the greatest extent possible, development of radicals and fragments from the volatilized hazardous organics, oxidation of the volatile radicals and fragments, and direct oxidation of chars resulting from certain volatilization reactions. These reactions are reviewed below.

1. *Volatilization of the Solids*

Volatilization is perhaps the most critical reaction to manipulate, as it largely determines the success of the subsequent incineration processes. Volatilization of the hazardous solid substance can occur by pyrolysis, melting and boiling, or sublimation. In the case of pyrolysis, the products are both gaseous compounds and char. In the cases of melting and boiling, and sublimation, the entire mass of solid is brought to the gaseous state.

a. Pyrolysis. Pyrolysis is the heating of any substance in the absence of sufficient oxygen to accomplish combustion. Pyrolysis may be initiated at relatively low temperatures, typically 500-650°F (260-340°C). For hazardous waste incineration, the pyrolysis mechanism is most prominent when applied to solid substances such as hexachlorobenzene or pesticides (i.e., aldrin, atrazine), or to chemicals absorbed onto a cellulosic material. The pyrolysis mechanism is generally reported as follows:

$$\text{Solid organic} + \text{heat} => CO_2 + H_2O + CO + CH_4$$

$$+ C_2H_6 + \ldots + \text{Char} \qquad (5\text{-}8)$$

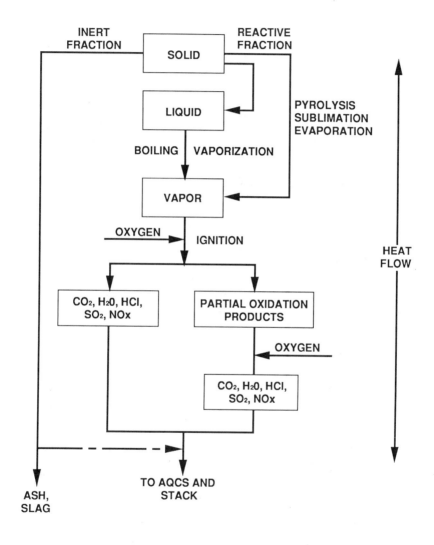

Figure 6. Global mechanism for the thermal oxidation of solid hazardous wastes.

The char is rich in fixed carbon, although it can contain hydrogen, oxygen, and other elements (see Chapter II). If halogens, sulfur, or nitrogen are present in the solid organic material being pyrolyzed, they will exhibit a strong tendency to report with the volatile fraction rather than the char fraction.

When applied to solid reactants, the pyrolysis mechanism generally yields most of the material as volatiles. In the cases of pesticide pyrolysis shown previously as TGA curves, the splits were typically 90-95% as volatiles and 5-10% as char. Cellulosic materials typically pyrolyze to 75-85% volatiles and 15-25% char [23]. The presence of volatile hazardous chemicals in a cellulosic material would increase the percentage going to volatiles into the 80-90% range. The pyrolysis mechanism can be manipulated by the application of temperature. As discussed in Chapter II, higher temperatures (and smaller particles) can be used to increase the proportion of material going to volatiles while lower temperatures will increase the fraction of the reporting as char.

   b. *Melting and Boiling, and Sublimation*. Melting and boiling, and sublimation are prominent mechanisms in the volatilization of some solid hazardous wastes, and in the generation of volatiles from soils, sediments, and inert sorbants contaminated with volatile and semi-volatile organics. Like pyrolysis, these are endothermic reactions. Further, they may occur at relatively low temperatures as shown in Table VII.

Melting and boiling, and sublimation, are heavily influenced by the soil or inert material matrix. Further, the matrix can influence the energy required to accomplish desorption and volatilization. Pershing and coworkers [13, 14, 29] have demonstrated that coarser soils such as sands are readily desorbed of volatile organics such as toluene and xylene. As soil particle sizes are reduced, energy required for desorption increases. Clays and silts require the most energy for desorption. Similar increases in the energy requirement for desorption may be caused by porous soil particles. Essentially increased difficulties occur in transferring heat to the hazardous waste within the soil matrix when soil particle sizes are reduced.

Solids devolatilization is significantly influenced by molecular weight and structure, and consequently boiling point, of the compounds being released from a soil matrix, as well as by the composition of the matrix itself. Work at the University of Utah [13, 14, 29] demonstrates that compounds will be released from a soils matrix by specific compound. When gasoline was thermally desorbed, toluene and octane were initially released. Subsequently, o-xylene

TABLE VII.   Melting and Boiling Points for Some
Hazardous Chemicals

Compound name	Melting point		Boiling point	
	°F	°C	°F	°C
Aniline	20.8	−6.2	363.2	184.4
Benzene	41.92	5.51	176.17	80.09
Carbon tetrachloride	−8.7	−22.6	170.2	76.8
1,2 dichlorobenzene	0.5	−17.5	356−361	180.183
Hexachlorobenzene	448	231	619	326
Nitrobenzene	43	6	412	210.9
Pyridine	44	42	239.5	115.3
1,2,4-trichlorobenzene	63	17	415	213
Atrazine	347	175	−	−
Aldrin	219	104	−	−
Chlordane	−		347	175
Dieldrin	302	150	347−349	175−176
Dicyclopentadiene	91.2	32.9	331.9	116.6
Hexachlorocyclopentadiene	49.8	9.9	462	239
Ethyl chloride	−217.7	−138.7	54.1	12.3
Methyl chloride	−143	−97	−10.7	−23.7
Methylene chloride	−141	−96	103.55	39.75
Chloroform	−82.3	−63.5	142.3	61.26
Chlorobenzene (phenyl chloride)	−49	−45	269.1	131.7
Phenol	105.53	40.85	359.4	181.9

was released. Finally, o-xylene and p-xylene were
desorbed.
     Pershing and coworkers [13, 14) also have demonstrated
that hazardous contaminants are chemisorbed onto the soil
matrix, causing much of the difficulty in achieving volatilization. Subsequently they have shown that melting and boiling,
and sublimation devolatilization mechanisms can be manipulated

by temperature, much as pyrolysis mechanisms are manipulated
by this parameter. Using clay as a substrate and xylene as
a contaminant, they have shown that higher temperatures
achieve more complete volatilization, and in a shorter residence time. Lower temperatures, with concomitantly lower
energy inputs, achieve reduced levels of volatile production
and also result in the formation of chars and soots as low
temperature pyrolysis mechanisms begin to compete with boiling
and sublimation mechanisms.

2. Gas Phase Reactions

Once volatiles are produced, they are subjected to gas
phase reactions such as those discussed in Chapter II. The
volatiles undergo additional decomposition reactions producing reactive radicals and fragments. These radicals and fragments then react with incoming oxygen (air) or chlorine within
the hazardous waste to form intermediate compounds and ultimately CO_2, H_2O, and HCl. The dominant gas phase decomposition reactions include OH addition to aromatic structures such
as hexachlorobenzene, other chlorinated benzenes, napthalene,
and vinyl chloride; OH abstraction reactions attacking such
compounds as chloromethane, chlorotoluene, formaldehyde, and
acetaldehyde; pyrolysis and unimolecular decomposition of such
compounds as chloroform, dinitrobenzene, and hydrazene; and
fragmentation or pyrolysis type reactions applied to such
compounds as nitroglycerine [27].
The reactions subsequent to degradation are analogous to
those discussed in Chapter II. The chain propagation and
branching reactions are dominated by hydroxy radical reactions,
as the OH radical is the most reactive specie in the combustion
process. The most prominent difference between thermal oxidation mechanisms associated with MSW and hazardous waste combustion is the potential for high concentrations of chlorine
and other halogens. Pyrolytic reactions breaking the C-Cl
bond can readily produce Cl radicals and stable intermediates.
Pyrolytic reaction sequences can include a wide range of pathways such as the following [21].

$$CCl_4 \Rightarrow C(s) + 2Cl_2 \qquad\qquad (5\text{-}9)$$

$$CH_2Cl_2 \Rightarrow C(s) + 2HCl \qquad\qquad (5\text{-}10)$$

$$CHCl_3 \Rightarrow C(s) + HCl + Cl_2 \qquad\qquad (5\text{-}11)$$

$$C_2H_5Cl \Rightarrow C_2H_4 + HCl \qquad\qquad (5\text{-}12)$$

Bond dissociation energies are relevant to each type of reaction (Table VIII). Noteworthy are the Cl- bond strengths which are substantially less than the H- bond strengths. This has obvious implications for the immediate products of pyrolysis.

After pyrolytic degradation, chlorine can react with any number of species, and produce a wide variety of compounds, during the combustion process. In the presence of stoichiometric or greater quantities of hydrogen, reactions will proceed along pathways leading to HCl. Hydrogen may enter the system with supplementary fuel (i.e., natural gas), with the

TABLE VIII. Bond Dissociation Energies of Compounds Relevant to CHC Combustion

Parent Complex	Bond strength (kcal/g-mol at 298°K)	
	H	Cl
H-	104	103
Cl-	103	58
CH_3-	104	84
C_2H_5-	98	81
C_3H_7-	95	81
C_4H_9-	92	81
C_6H_5-	110	95
$C_6H_5CH_2-$	85	69
CCl_3-	96	73
CH_3CO-	86	82
CN-	120	97
OH-	119	60
C_6H_5CO-	-	74
CH_2Cl-	101	-
$CHCl_2-$	96	-
C_2HCl_4-	94	-
$CH_2=CH-$	108	91

Source: Senkan [21].

hazardous waste (i.e., chlorobenzene), or with moisture. Representative reactions would include the following [21]:

$$C_2(g) + Cl_2 => 2CCl(g) \tag{5-13}$$

$$CCl + H_2O => HCO + HCl \tag{5-14}$$

With HCO being a reactive species that will ultimately go to H_2O and CO_2. Representative reactions also can include:

$$Cl_2 + O_2 => 2ClO \tag{5-15}$$

$$H_2O + ClO => HOCl + OH \tag{5-16}$$

A wide range of other reactions is equally possible for this system. All such reactions ultimately lead to the generation of HCl as the preferred product.

In the absence of hydrogen, however, chlorine atoms can react in a wide variety of even less desirable pathways. For example, Cl_2 can be generated, and it can react with CO to form phosgene gas or carbonyl monochloride by the following pathways [21]:

$$CO + Cl_2 => COCl_2 \tag{5-17}$$

$$CO + 0.5Cl_2 =>' COCl \tag{5-18}$$

In high temperature systems, however, CO is favored over phosgene or carbonyl monochloride. However phosgene has been produced in the presence of iron when CCl_4 is oxidized. Other compounds yielding phosgene include formaldehyde, trichlorethylene, dichlorethylene, and perchloroethylene [21]. Chlorine can also promote the formation of soot in systems utilizing low excess air. Further, chlorine can promote the formation of polyaromatic hydrocarbon (PAH) compounds under certain conditions.

The influence of chlorine on the combustion system, then, is quite complex. However, in the presence of sufficient hydrogen, HCl is the dominant product. In the absence of sufficient hydrogen, less desirable end products may result. Further, chlorine can increase the formation of soot and PAH compounds under certain circumstances.

3. *Gas-Solids Reactions*

Pyrolysis of solid wastes produces not only volatiles, but also carbon-rich chars, as previously noted. These chars contain largely carbon, but also can contain heteroatoms such

as hydrogen, oxygen, nitrogen, and sulfur. Gas-solids oxidation occurs by a number of mechanisms as discussed in Chapter II. These mechanisms include the steam-carbon reaction
producing CO and H_2, which ultimately burns in the gas phase;
the Boudouard reacton generating CO for subsequent gas phase
oxidation; and, the direct oxidation of reactive carbon sites
in the char. All of these reactions occur as solid carbon is
burned to completion.

B. Temperature and Hazardous Waste Combustion

Two temperature issues must be addressed with solid
hazardous waste combustion: (1) the temperatures required for
combustion in order to meet regulatory requirements, and (2)
the flame temperatures generated by the combustion of solid
organic wastes. Both issues are considered below. Additionally, the practical consequences of temperatures are largely
specific to incineration system designs (i.e., rotary kiln,
fluidized bed) but are identified here as extensions of the
temperature issue.

1. *Temperatures Required for Hazardous Waste Incineration*

RCRA mandates that all hazardous wastes incinerated will
be destroyed to a DRE of 99.99%, while TSCA states that PCB
wastes will be destroyed to a DRE of 99.9999%. These regulations explicitly or implicitly impose temperature requirements
on the incineration system. For example, the RCRA temperature requirements are implicit in the 99.99% DRE, and are
readily calculated. As Tsang and Shabu [27] demonstrate,
equation 5-7 can be rearranged as follows:

$$\int_{Hz_i}^{Hz_f} -dHz/Hz = \ln Hz_i/Hz_f = \int_{o}^{r} [k_{uni}$$

$$+ (\Sigma k_{bi}R_i)] \, dt \qquad (5-19)$$

Where r is residence time in the incinerator. Given a DRE of
99.99%, the log term in the equation is 9.212. Solving for
the temperature requirement involves finding conditions where
9.212 is exceeded when values are determined for k_{uni}, k_{bi},
R_i, and T. Since the rate constants are obtained independently of the incineration regime, and since R_i defines the
waste being incinerated, this equation is solved for T [27].
Such temperatures and incineration conditions assume that

the reactant is in the gas phase. Temperatures required to pyrolyze, boil, or sublime the solids and produce gaseous molecules range from 500–1800°F (260–982°C), depending upon the material being reacted as shown previously.

As a practical matter, residence times can be set at 1 to 2 seconds because rate constants are well known. Consequently, temperatures required for incineration can be calculated for any given compound or mix of compounds. Data for calculating the temperatures required for incineration of a few representative compounds are shown in Tables VIII, IX, and X.

2. Temperatures Achieved During Solid Hazardous Waste Combustion

Temperatures achieved during solid hazardous waste combustion vary over the widest possible range due to the compositions available. Theoretical flame temperatures can be calculated by the general equation below:

$$\Sigma H_p + \Sigma n_i \int_{298°K}^{T} C_{p_i} \, dT + \Sigma n_i \, \lambda_i \qquad (5-20)$$

Where H_p is enthalpy of products, C_p is molal heat capacity, T is absolute temperature, and λ is heats associated with phase changes. For detailed calculations, the Gordon-McBride model, discussed in Chapter II [9], provides probably the most accurate method of evaluation [2,6].

Adiabatic and actual flame temperatures can be calculated and estimated for a wide variety of hazardous compounds. Substantial variation can occur as a function of incinerator regime (i.e., use of supplementary fuel, excess air, air preheat, and related factors). Variation also can occur as a function of waste composition with particular emphasis on: (1) ultimate analysis, (2) higher heating value, (3) chlorine content and the expected function of chlorine (as an oxidant), (4) moisture content, and (5) inerts content. Moisture and inerts are particularly influenced by the matrix which may hold the hazardous waste. As can be seen from the equations, however, organic hazardous wastes bound in a soil matrix can only be incinerated with substantial use of supplementary fuels.

3. Problems Associated with Hazardous Waste Combustion Temperature

Temperature is a regulatory issue which must be addressed for solids based reactors such as rotary kilns (not including

TABLE IX.   Temperature (°K) Necessary to Assure 99.99
Percent Destruction Assuming Unimolecular Mechanism
and Arrhenius Rate Expression ($k = A_{oo}exp\ (-E_{oo}/RT)$)
for Incerator Residence Time of Approximately 1 Second

A_{oo}	E_{oo} (kcal)							
(/sec)	40	50	60	70	80	90	100	110
10^{12}	792	990	1,188	1,386	1,584	1,782	1,980	2,178
10^{13}	726	908	1,089	1,271	1,453	1,634	1,815	1,980
10^{14}	670	838	1,006	1,173	1,341	1,509	1,676	1,844
10^{15}	622	778	934	1,090	1,245	1,401	1,556	1,713
10^{16}	581	727	872	1,017	1,163	1,308	1,453	1,599
10^{17}	545	681	818	954	1,090	1,226	1,362	1,499

Source: Tsang and Shaub [27].

TABLE X.   Rate Expressions (/sec) for the Decomposition
of Alkyl Chlorides

$C_2H_5Cl - C_2H_4 + HCl$	$10^{13.45}exp\ (-55,000/RT)$
$nC_3H_7Cl - C_2H_6 + HCl$	$10^{13.6}exp\ (-55,150/RT)$
$nC_4H_9Cl - C_4H_8 + HCl$	$10^{13.2}exp\ (-56,500/RT)$
$iC_3H_8Cl - C_3H_6 + HCl$	$10^{13.16}exp\ (-51,100/RT)$
$sC_4H_9Cl - C_4H_8 + HCl$	$10^{14.0}exp\ (-50,600/RT)$
$tC_4H_9Cl - C_4H_8 + HCl$	$10^{13.8}exp\ (-45,000/RT)$
$tC_5H_9Cl - C_5H_{10} + CHl$	$10^{13.7}exp\ (-44,000/RT)$

Source: Tsang and Shaub [27].

afterburners), fluidized beds, and other systems. Increasing
the temperature in the primary reactor decreases the residence
time of gaseous reactants in that reactor [13, 14]. However,
this increases the burden on the secondary reactor in a two-
stage system [20, 25]. Temperature increases in the primary
reactor also create the potential for "puffs," or occasions
when the pressure in the primary reactor becomes positive and
an instantaneous, sharp increase in load is experienced by the
secondary reactor [30]. Temperature also may have deleterious
impacts on the behavior of the solids matrix (depending upon
design), and upon the formation of certain airborne emissions
(i.e., NO_x).

Low temperatures can cause a different set of equally
vexing problems. As can be inferred from the discussion below,
low temperature can cause inadequate volatilization of the
contaminants, leading to inadequate treatment of the materials.
Further, low temperatures, if caused by excess air problems,
can lead to excessive emissions of carbon monoxide, hydrocarbons, and potentially dioxins and furans [26, 12]. Much of
the discussion regarding control of the temperature variable
is carried in Chapters VI and VII. The influence of temperature on airborne emission formation is considered in the subsequent portion of this chapter.

C.   Formation of Airborne Emissions from Hazardous
     Waste Combustion

Airborne emissions critical to any installation are:
(1) particulates, (2) sulfur dioxide (SO_2), (3) hydrogen
chloride (HCl), (4) oxides of nitrogen (NO_x), (5) carbon
monoxide (CO) and hydrocarbons (HCs), and (6) dioxins and
furans. Particulate formation is a function of specific reactor design and operation, as well as the composition of the
waste (particularly the inert content of the waste). Acid
gases such as SO_2 and HCl are formed strictly as a function
of the waste composition and the complete oxidation of sulfur
(by oxygen) and hydrogen (by chlorine). Important combustion
mechanisms, then, are those associated with formation of NO_x
and organic emissions (CO, HCs, and dioxins and furans).

1. *NO_x Emission Formation from Hazardous Waste Incineration*

As discussed in Chapter II, there are two sources of
oxides of nitrogen: (1) thermal NO_x and (2) fuel NO_x. Both
of these sources are potentially important in hazardous waste
incineration.

a. *Thermal NO$_x$.* Thermal NO$_x$ is typically formed at adiabatic flame temperatures exceeding 3200°F, or actual flame temperatures exceeding 2700°F [17], and is formed by a complex sequence of mechanisms. The dominant mechanism for thermal NO$_x$ formation was elucidated by Zeldovich [16] as follows:

$$O + N_2 \Rightarrow NO + N \qquad\qquad (5-21)$$

$$N + O_2 \Rightarrow NO + O \qquad\qquad (5-22)$$

Other thermal NO$_x$ reactions have been reviewed in detail by Glassman [8] and include:

$$N + OH \Rightarrow NO + H_2 \qquad\qquad (5-23)$$

$$CH + N_2 \Rightarrow CHN + N \qquad\qquad (5-24)$$

$$CN + H_2 \Rightarrow HCN + H \qquad\qquad (5-25)$$

$$CN + H_2O \Rightarrow HCN + OH \qquad\qquad (5-26)$$

HCN then goes to NO according to the prompt NO formation mechanism suggested by Fenimore [8]. These reactions have been reviewed in Chapter II. What is important to solid hazardous waste incineration is not the specific mechanism per se, but the fact that thermal NO$_x$ can be formed in appreciable quantities at actual flame temperatures exceeding 2700°F (1480°C). Since high temperatures are required for devolatilization of many soil matrices, and since regulations periodically require exit gas temperatures from incineration systems on the order of 2200°F (1200°C), flame temperatures in excess of 2700°F (1480°C) are possible in all but the fluidized bed hazardous waste incineration systems. Consequently thermal NO$_x$ can be generated in concentrations of several hundred parts per million by volume of stack gas (ppmv), depending upon the explicit combustion regime, as has been measured at several incinerator locations.

b. *Fuel NO$_x$.* Fuel NO$_x$, generated by the oxidation of nitrogen entering as part of the waste or supplementary fuel source, can also be a problem with solid hazardous wastes. Fuel nitrogen typically is of minor consequence with the supplementary fuels. Typically, natural gas and No. 2 distillate oil are used as supplementary fuels, and they are very low in nitrogen content. Nitrogen can come in with the hazardous waste, however, particularly in compounds such as analine, pyridine, nitrile compounds, nitrobenzene, pesticides, and related materials.

Fuel or waste bound nitrogen among hazardous wastes has been shown previously. As can be seen, hazardous wastes are not typically highly nitrogenous. Consequently the fuel NO_x mechanisms are not as significant as the thermal NO_x mechanisms when solid hazardous waste is incinerated, except such wastes as pesticides are being burned. This situation is in direct contrast to the municipal waste NO_x formation and control problem as discussed in Chapter II.

2. Organic Emissions

Organic emissions include CO, HCs, and dioxins and furans and are of particular concern in hazardous waste incineration. Carbon monoxide and hydrocarbons are the best indicator of organic emissions that can be continuously monitored. Hydrocarbons often are identified as products of incomplete combustion (PICs). Dioxins and furans have health risk consequences well identified in the literature.

a. CO and HCs. Carbon monoxide and hydrocarbons are caused by incomplete combustion--commonly as a result of insufficient oxygen availability to the radicals generated by pyrolysis and devolatilization, and subsequent gas phase decomposition reactions. This can be caused by insufficient excess air in the combustion system. Insufficient oxygen availability to the hydrocarbon radicals and fragments also can result from insufficient mixing in the combustion chamber. Carbon monoxide and hydrocarbon emissions can be formed by a combustion zone which is sufficiently cool that combustion reactions "freeze," and do not go to completion. Insufficient temperature for complete combustion typically is caused by large quantities of excess air cooling the reactor. CO and HCs are readily controlled by careful attention to the quantity and distribution of excess air. Rotary kiln systems are optimized when excess air is maintained at 35-50% [12]). This includes inleakage around kiln seals. Lower levels of excess air result in incomplete combustion. Higher levels of excess air result in excess CO caused by cool combustion. The problem of inleakage around kiln seals is of some consequence, however (see Fig. 7), such inleakage makes excess air levels below 50% difficult to maintain.

b. Dioxins and Furans. Dioxin and furan emissions can be formed by a variety of mechanisms (see Chapter II). Dominant among these mechanisms are the formation of 2,3,7,8 TCDD

FIGURE 7. Closeup of the interface between the front
wall and rotating portion of a slagging rotary kiln.
This interface is a primary source of kiln inleakage
of air. It limits the lower bound of excess air achievable in kiln operation. (Photo courtesy of Environmental
Elements Co.)

and 2,3,7,8 TCDF from chlorinated precursors such as PCB
oils, chlorinated phenols, hexachlorobenzene, and related
compounds.

Shaub and Tsang [22] performed extensive kinetic studies
on alternative cases for hazardous waste incineration. In an
analysis of incineration of a pure chlorophenol, very low
concentrations of TCDD and TCDF could be estimated even when
the analysis was biased to calculate maximum dioxin and furan
formation (Table XI). Similarly, Scrivner [19] has demonstrated that, if dioxins are formed during the combustion
process, they will be destroyed to greater than 99.99% DRE at
temperatures exceeding 2000°F (1093°C) in residence times of
less than 0.1 second. A degradation curve for hexachlorobenzene and other potential dioxin precursors is shown in
Fig. 8 [22].

Clearly virtually all chlorinated dioxins and furans are
readily destroyed, if formed, by properly performed high
temperature hazardous waste incineration; only minute

TABLE XI.   Calculated Concentration of Dioxins Formed
    From Incineration of a Hazardous Waste (e.g.,
    Chlorophenol) Under Oxidizing Conditions as a
    Function of Temperature

Temperature		Dioxin concentration (mol/1)	Chlorinated dioxin precursor concentration (mol/1)
(°K)	(°F)		
500	440	7.95×10^{-29}	8.58×10^{-23}
600	620	3.69×10^{-21}	2.05×10^{-18}
700	800	8.43×10^{-16}	1.85×10^{-15}
800	980	5.64×10^{-13}	1.23×10^{-14}
900	1160	3.48×10^{-11}	3.02×10^{-14}
1000	1340	6.68×10^{-10}	3.23×10^{-14}
1100	1520	2.29×10^{-9}	1.91×10^{-15}
1200	1700	3.61×10^{-12}	8.10×10^{-21}
1300	1880	7.97×10^{-33}	2.19×10^{-37}
1400	2060	$<1.00 \times 10^{-70}$	$<1.00 \times 10^{-70}$
1500	2240	$<1.00 \times 10^{-70}$	$<1.00 \times 10^{-70}$

Source: Shaub and Tsang [22].

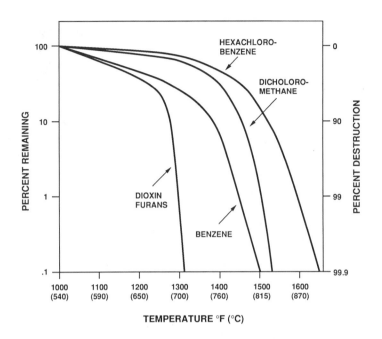

FIGURE 8. Degradation curves for dioxins and furans, and selected additional hazardous waste compounds. Source: [22].

quantities remain. Further, high temperatures and sufficient residence times will produce sufficient degradation of precursors into increasingly small, readily oxidized fragments so that postcombustion formation of dioxins (i.e., by condensation on particulates) is less likely. Like CO and HC emissions, then, dioxin and furan emissions are minimized by the application of good combustion practice.

3. *Heavy Metals Emissions*

Heavy metals are the final area of concern in this chapter. Emissions of heavy metals have come under increased scrutiny due to their potential for deleterious health effects. While such emissions are of concern in the combustion of municipal solid waste, they are of equal or greater importance in the incineration of hazardous wastes. This

importance stems from the concentrations of such metals as
lead, mercury, cadmium, chromium, and other toxic inorganics
in potential feed streams to incinerators. For example, the
soils at Bog Creek Farm contained up to 5,000 ppm (0.5%) lead.

The fate of heavy metals in hazardous waste incinerators
depends upon the metal and its particular physical and thermodynamic characteristics. At different temperatures, various
metals will vaporize. In reality a gradient will be established indicating that, for any given temperature, some proportion of the metal will enter the vapor state. In the
process, the metal may be oxidized as well. After the metal
is vaporized, the ion leaves the incinerator in the gaseous
products of combustion. As the gas cools, various metals will
condense to greater or lesser extents. This condensation is
particularly prevalent on fine particulate, where there is
extensive surface area available for condensation. In summary,
some portion of the metal typically does not vaporize, but
remains in the solid products of combustion. The remaining
portion of the metal vaporizes. A fraction of the vaporized
metal will recondense while the remainder will escape the
particulate collection system in the gas phase.

The extent to which metals vaporize is temperature dependent. Very hot incinerators (i.e., slagging rotary kilns)
will experience more metal vaporization than moderate temperature incineration systems such as ashing rotary kilns and
fluidized beds. Similarly the recondensation phenomenon is
temperature dependent. As the air quality control system
temperature is reduced, more metal will condense out either in
the system or on the particulate. Recent advances in condensing and ionizing wet scrubbers, and low temperature dry
injection systems, have had significant success in capturing
volatilized metals by promoting condensation (see Chapter
VIII).

For all incineration systems except slagging rotary kilns,
the temperature of the solid reactants will typically be in
the 1300°F-1500°F (704°C-815°C) region. Seeker et al. [20]
have performed theoretical calculations to indicate the fate
of selected heavy metals at 1500°F (815°C), both in terms of
degree of vaporization and species expected. Further,
Seeker et al [20] have calculated the extent and nature of
condensation for such metals in a typical air quality control
system. These calculations are shown in Tables XII an XIII.
Such calculations provide a basis for evaluating the issues
associated with air quality control system design and
operation.

IV. CONCLUSION: CONSEQUENCES OF FUEL PROPERTIES AND
     PRINCIPLES OF HAZARDOUS WASTE COMBUSTION FOR
     INCINERATION SYSTEMS

     Incineration is frequently proposed and implemented as a
hazardous waste treatment and disposal option. Fuel properties
and principles of combustion provide insights into the best
method of incinerating hazardous waste.
     Fuel properties impacting incineration system selection,
design, and operation include the type of solids to be treated,
their ultimate analysis and higher heating value, their halogen content, and the matrix in which the contaminants exist.
These variables impact fuel reactivity, the need for supplementary fuels such as distillate oil and natural gas, and the
combustion regime most appropriate to achieve the required
DRE level. All fuel properties, including ash fusion temperature and related characteristics, impact the selection of the
most appropriate incineration system.
     The combustion mechanism itself, when applied to hazardous
wastes, involves a special case of the three "T's" of combustion: time, temperature, and turbulence. It is an application of those principles to a material that may be very high
in halogen content, ash/inerts content, and water content.
It is an application of those principles where legal regulations often dictate combustion regime, and consequently the
way in which the combustion process is manipulated. Within
these constraints, however, the combustion mechanism as
applied to hazardous wastes can be manipulated by regime
selection to maximize the devolatilization of the solid material and to make most likely the removal of contaminants from
any solid matrix. Further, the mechanism can be manipulated
for maximum destruction of POHCs while minimizing formation
of organic pollutants. Manipulation of the combustion mechanism involves hardware selection and design. Such issues
are the focus of Chapters VI and VII.

TABLE XII. Theoretical Partitioning of Selected Heavy Metals in a Rotary Kiln Environment Operating at 1500°F (815°C) with Small Entrained Particles: No Chlorine in Feed

	Product Distribution					
	Residual		Fly ash[a]		Escaping	
Metal	Fraction	Species	Fraction	Species	Fraction	Species
As	0.0	–	0.005	As	0.0	As
Ba	0.892	BaO	0.047	BaO	0.061	BaO
Be	0.950	BeO	0.0498	BeO	0.0002	BeO
Cd	0.161	CdO	0.0127	CdO	0.826	CdO
Cr	0.950	$NiCr_2O_4$	0.0498	$NiCr_2O_4$	0.0002	$NiCr_2O_4$
Cu	0.950	Cu	0.0498	Cu	0.0002	Cu
Pb	0.0	–	0.006	Pb	0.994	Pb
Hg	0.0	–	0.0	–	1.0	Hg
Ni	0.950	$Ni,NiCr_2O_4$	0.0498	$Ni,NiCr_2O_4$	0.0002	$Ni,NiCr_2O_4$
Ag	0.950	Ag	0.0498	Ag	0.0002	Ag
V	0.950	V_2O_3	0.0498	V_2O_3	0.002	V_2O_3

[a]Small particles (peak at 5 microns); Venturi scrubber.

Source: Seeker et al. [20].

194

TABLE XIII. Theoretical Partitioning of Selected Heavy Metals in a Rotary Kiln Environment Operating at 1500°F (815°C) with Small Entrained Particles: 10% Chlorine in Feed

Product Distribution

Metal	Residual Fraction	Residual Species	Fly ash[a] Fraction	Fly ash[a] Species	Escaping Fraction	Escaping Species
As	0.0	-	0.005	As	0.995	As
Ba	0.950	$BaCl_2$	0.0498	$BaCl_2$	0.0002	$BaCl_2$
Be	0.950	BeO	0.0498	BeO	0.0002	BeO
Cd	0.161	CdO	0.0127	CdO	0.820	CdO
Cr	0.950	Cr_2O_3	0.0498	Cr_2O_3	0.0002	Cr_2O_3
Cu	0.0	-	0.006	Cu	0.994	Cu
Pb	0.0	-	0.006	Pb	0.994	Pb
Hg	0.0	-	0.0	-	1.0	Hg
Ni	0.909	Ni	0.0479	Ni	0.0431	Ni
Ag	0.855	Ag	0.00994	Ag	0.0954	Ag
V	0.950	V_2O_3	0.0498	V_2O_3	0.0002	V_2O_3

[a]Small particles (peak at 5 microns); Venturi scrubber.

Source: Seeker et al. [20].

REFERENCES

1. Cegielski, J. M. 1981. Hazardous Waste Disposal by
    Thermal Oxidation. John Zink Co., Tulsa, Oklahoma.

2. Diemer, R. B., Ellis, T. D., Vevai, J. E. Scrivner, N. C.,
    and Brunner, C. R. 1987. Hazardous Waste Incineration.
    In The Encyclopedia of Physical Science and Technology,
    Vol. 6. Academic Press, San Diego, California, PP. 436-
    450.

3. Dellinger, B. et al. 1984. Determination of the Thermal
    Decomposition Properties of 20 Selected Hazardous Organic
    Compounds. University of Dayton Research Institute,
    Dayton, Ohio. For USEPA Incineration Research Branch,
    Industrial Environmental Research Laboratory, Cincinnati,
    Ohio.

4. Dellinger, B. 1987. Thermal Stability Based Ranking of
    Hazardous Organic Compound Instability. Proc: American
    Flame Research Committee International Symposium on
    Incineration of Hazardous, Municipal, and Other Wastes,
    Palm Springs, California.

5. Ebasco Services Incorporated. 1988. Prefinal Design
    95% Submittal Task D9 - Remedial Design: Bog Creek Farm
    Site, Township of Howell, Monmouth County, New Jersey.
    Vol. II of II.

6. Ellis, T. D., Scrivner, N. C., Vevai, J. E., Brunner,
    C. R. 1983. Industrial Hazardous Waste Incineration
    (Syllabus for AIChE Today Series short course). American
    Institute of Chemical Engineers, New York. Also see:
    Scrivner, N.C. 1983. Thermodynamics and Kinetics of
    Dioxin Destruction. Hazardous Waste Incineration, AIChE
    Today Series. Supplement #2.

7. Ferguson, T. L. et al. 1975. Determination of Incinera-
    tor Operating Conditions Necessary for Safe Disposal of
    Pesticides. Prepared by Midwest Research Institute for
    Municipal Environmental Research Laboratory, USEPA.
    Contract No. 68-03-0286.

8. Glassman, I. 1987. Combustion, 2nd Edition. Academic
    Press, San Diego, California.

9. Gordon, S. and McBride, B. 1971. Computer Program for
    Calculation of Complex Chemical Equilibrium Concentra-
    tions. National Aeronautics and Space Administration,
    Washington, D.C.

10. Hazen Research. 1987. Results of Laboratory Tests in Support of Bog Creek Farm Incineration Program. REM-III CERCLA (Superfund) Site.

11. Hougen, O. F., Watson, K. A., and Ragatz, R. A. 1954 Chemical Process Principles, Part 1: Material and Energy Balances (2nd Ed.). John Wiley & Sons, New York.

12. Kramlich, J. C. et al. 1984. Laboratory-Scale Flame-Mode Hazardous Waste Thermal Destruction Research. EER Corp., Irvine, California for USEPA Contract EPA-600/2-84-086.

13. Lighty, J. S., Silcox, G. D., Owens, W. D., Pershing, D. W., and Cundy, V. A. 1987. Fundamentals of Hazardous Solid Waste Incineration in a Rotary Kiln Environment. Proc: American Flame Research Committee International Symposium on Incineration of Hazardous, Municipal, and Other Wastes. Palm Springs, California.

14. Lighty, J. S., Owens, W., and Pershing, D. W. 1987. Utilization of Natural Gas for Incineration Processes. Monthly Report: March, 1987. Department of Chemical Engineering, University of Utah, Salt Lake City, Utah for Gas Research Institute, Chicago. Also: Lighty, J.S., Van Os, R., and Pershing, D. W. 1987. Utilization of Natural Gas for Incineration Processes. Monthly Report: April, 1987. Also: Lighty, J. S., Owens, W., Van Os, R. and Pershing, D. W. 1987. Utilization of Natural Gas for Incineration Processes. Monthly Report: May, 1987. Lighty, J. S., and Pershing, D. W. 1987. Utilization of Natural Gas for Incineration Processes. Monthly Report: June 1987.

15. Linak, W. P., McSorley, J. A., Wendt, J. O., and Dunn, J. E. 1987. On the Occurrence of Transient Puffs in a Rotary Kiln Incinerator Simulator. JAPCA 37(8):934-942.

16. Palmer, H. B. 1974. Equilibrium and Chemical Kinetics in Flames. In Combustion Technology: Some Modern Developments (H. B. Palmer and J. M. Beer, eds.). Academic Press, New York.

17. Pershing, D. W., and Wendt, J. 1976. Pulvarized Coal Combustion: The Influence of Flame Temperature and Coal Composition on Thermal and Fuel NO_x. Proc: Sixteenth International Sympoisum on Combustion. The Combustion Institute. Also see: Pershing, D. W. et al. 1978. The Influence of Fuel Composition and Flame Temperature on the Formation of Thermal and Fuel NO_x in Residual Oil Flames. Proc. Seventeenth International Symposium on Combustion. The Combustion Institute.

18. Santolari, J. 1986. Incineration Technology: State-of-the-Art Review. Permitting Hazardous Waste Incinerators Short Course. USEPA, San Francisco, California.

19. Seeker, W. R., Lanier, W. S., and Heap, M. P. 1987. Combustion Control of MSW Incinerators to Minimize Emissions of Trace Organics. EER Corporation, Irvine, California for USEPA under Contract No. 68-02-4247.

20. Seeker, W. R. et al. 1987. The Characterization of the incinerability of Waste Samples from Bog Creek Super-fund Site. EER Corporation, Irvine, California, for Envirosphere Co. in Support of REM-III Assignment.

21. Senkan, S. 1982. Combustion Characteristics of Chlorinated Hydrocarbons. *In* Detoxication of Hazardous Waste (J. H. Exner, ed.). Ann Arbor Science, Ann Arbor, Michigan.

22. Shaub, W. M., and Tsang, W. 1983. Dioxon Formation in Incinerators. *Environmental Science and Technology 17* (12):721-730.

23. Tillman, D. A. 1987. Biomass Combustion. *In* Biomass: Regenerable Energy (D. O. Hall, and R. P. Overend, eds.) John Wiley & Sons Ltd., London, pp. 203-219.

24. Tillman, D. A., Rossi, A. J., and Kitto, W. D. 1981. Wood Combustion: Principles, Processes, and Economics. Academic Press, New York.

25. Tillman, D. A., Seeker, W. R., Pershing, D. W., and DiAntonio, K. 1988. Converting Treatibility Tests into Conceptual Incineration Designs. Presented at The Combustion Institute Western States Meeting, Salt Lake City, Utah.

26. Trendholm, A., Gorman, P., and Jungclaus, G. 1984. Performance Evaluation of Full-Scale Hazardous Waste Incinerators. Midwest Research Institute, Kansas City, Missouri for USEPA. Contract No. 68-02-3177.

27. Tsang, W., and Shaub, W. M. 1982. Chemical Processes in the Incineration of Hazardous Materials. *In* Detoxification of Hazardous Waste (J. H. Exner, ed.). Ann Arbor Science, Ann Arbor, Michigan.

28. Tsang, W. 1987. High Temperature Chemical and Thermal Stability of Chlorinated Benzenes. Proc: American Flame Research Committee International Symposium on Incineration of Hazardous, Municipal, and Other Wastes. Palm Springs, California.

29. Van Os, L. M., Lighty, J. S., Pershing, D. W., and Cundy, V. A. 1988. Thermal Incineration as a Method of Contaminated Soil Clean-up. Presented at the Combustion Institute Western States Meeting, Salt Lake City, Utah.

30. Wendt, J. O. L., Linak, W. P., and McSorley, J. A. 1987. Fundamental Mechanisms Governing Transients From the Batch Incineration of Liquid and Solid Wastes in Rotary Kilns. Proc: American Flame Research Committee International Symposium on Incineration of Hazardous, Municipal, and Other Wastes. Palm Springs, California.

Chapter VI

PERMANENT SOLID HAZARDOUS WASTE

INCINERATION SYSTEMS

I. INTRODUCTION

Hazardous waste incineration systems have been designed
to capitalize on the thermal oxidation mechanisms discussed
in Chapter V. Because of the ability of incineration to
chemically transform hydrocarbon and halocarbon type materials
into CO_2, HCl, and H_2O, incineration is often considered to
be an "ultimate" disposal approach to hazardous solids. Consequently, there are over 100 general and special purpose
solid hazardous waste incinerators operating in the United
States [8], and a substantial additional number under
construction.

A. The Purpose of Solid Hazardous Waste Incineration

Hazardous waste incineration systems have a single purpose: to destroy materials legally defined as hazardous so
that they no longer present a long-term legal and financial
liability to generators. The purpose is legal exposure
reduction more than volume reduction. Energy recovery is
sometimes emphasized in Europe, but it is not important in
U.S. installations.

Hazardous waste incineration is governed largely by federal
law, as outlined in Chapter I. For wastes considered hazardous under the Resource Conservation and Recovery Act (RCRA),
the destruction and removal efficiency (DRE) required of an
incinerator is 99.99%, as discussed in Chapter V. The DRE
calculation considers the removal required for hazardous
wastes or principal organic hazardous constituents (POHCs)
which have been volatilized from the solid waste. For POHCs
which remain in the solid residues of incineration (bottom
ash or collected flyash), the options are to demonstrate sufficient cleanliness through a risk assessment, TCLP, and
EP-Toxicity analyses in order to delist the solids and consider them no longer hazardous; or to dispose of such solids
in a secure (hazardous waste) landfill. While RCRA requires

incinerators to operate at 99.99% DRE and to minimize the
formation of products of incomplete combustion (PICs); the
Toxic Substances Control Act (TSCA) governs the disposal of
polychlorinated biphenyls (PCBs), and it requires DRE levels
of 99.9999%. TSCA also may govern the actual combustion
regime (temperature, residence time, excess air) of the incinerator, as discussed in Chapter V.

Other federal laws governing hazardous waste include
Superfund (CERCLA, with SARA Amendments), and the Defense
Establishment Restoration Act (DERA) which governs military
installations. While specific DRE levels are not specified
in either Superfund or DERA regulations, incinerators
installed under such act must conform to the basic requirements of RCRA and TSCA. All incinerators also must conform
to all applicable state, regional, and local air quality,
water quality, and other applicable environmental regulations.

B. Types of Hazardous Waste Incineration Systems

Numerous incineration technologies have been developed to
incinerate the vast array of hazardous wastes produced as a
consequence of industrial economic activity. Many of these
techniques are employed at general purpose permanent facilities: those regional incinerators which serve customers by
accepting and destroying any appropriate wastes. Special purpose permanent incinerators typically are operated by manufacturers for the disposal of wastes generated by the owner.
Many manufacturers such as Dow Chemical, E. I. DuPont, Eastman
Kodak, and General Electric, own their own, permanently
installed, hazardous waste incinerators.

In order to meet the wide technical requirements of general
purpose incineration, technical approaches have been developed
that include rotary kilns, hearth type furnaces, fluidized
beds, and an array of emerging or experimental technologies
such as low temperature volatilization and pyrolysis. The
hazardous waste industry is gravitating towards rotary kiln
and fluidized bed systems for most solid hazardous waste incineration installations. Adjuncts to these dedicated technologies include cement kilns, aggregate kilns, pulp mill lime
kilns, and high efficiency industrial boilers retrofitted to
burn select hazardous wastes as supplementary fuel [12, 13,
21]. All of these technologies meet 99.99% DRE. Oppelt [12],
for example, shows that commercial rotary kiln installations
can achieve 99.9993% DRE on a variety of wastes, while high
efficiency boilers achieve 99.995 to 99.9997% DRE and cement
kilns can achieve as high as 99.992% DRE. Permanently
installed incineration systems, then, offer a potentially
successful method for destroying hazardous wastes. Specific

permanent solid hazardous waste incineration technologies of
particular interest here include rotary kiln and fluidized
bed designs. Current technology for incineration of solid
hazardous wastes focuses upon rotary kiln systems, with some
emphasis on the commercially emerging fluidized bed technologies.

C. Approach of This Analysis

This chapter focuses on combustion principles applied to
permanent, general purpose hazardous waste incinerators with
commentary on special purpose thermal destruction units. The
focus is the incinerator reactor system itself (i.e., the
kiln, the fluidized bed). Other subsystems such as materials
handling, solids treatment and disposal, wastewater and
process water systems, utilities, and balance of plant are
outside the scope of this analysis. The focus of analysis is
on the application of combustion principles and mechanisms,
along with the applications and limits of each type of combustion system.

II. ROTARY KILN INCINERATION SYSTEMS

Numerous permanent, general purpose hazardous waste incinerators exist throughout the United States and Europe.
Such incinerators in the U.S. include the installations built
and operated by Rollins Environmental Services in Bridgeport,
New Jersey, Baton Rouge, Louisiana, and Deer Park, Texas; the
facility now owned and operated by Waste Management, Inc. (SCA)
in Chicago, Illinois, the joint venture facility of National
Electric Company and Westinghouse in Coffeyville, Kansas, and
other similar installations. In western Europe, centralized
general purpose incinerators include the Kommunekemi facility
in Nyborg, Denmark, which was installed in 1975 and is currently being expanded; the SAKAB AB facility in Nortorrp,
Sweden, which began operations in 1984; and a host of other
plants. In Europe, general purpose incineration is a technology of choice for destroying hazardous wastes, particularly
those toxics being generated on an ongoing basis [16]. Recent
ongoing expansions of such facilities are shown here in
Table I. Furthermore, the popularity of solid hazardous
waste incineration is expected to grow dramatically in the
near future in the U.S. as the EPA ban on land disposal of
waste drives more hazardous material to thermal treatment
[15].

TABLE I. Hazardous Waste Incinerator Expansions
In Europe

Company	Location	Capacity (tonne/y)[a]	Startup year
AVR-Chemie	Rotterdam	40,000	1987
TREDI	St. Vulbas, France	6,000	1988
TREDI	Salaise, France	40,000	n/a
Rechem	Pontypool, U.K.	13,000	1987
ZSM	Schwabach, W. Germany	30,000	1988
SARP Ind.	Limay, France	50,000	1988
	Bassens, France	20,000	1988
SIDIBEX	Sandouville, France	35,000	1988
Kommunekemi	Nyborg, Denmark	50,000	1989
ABR	Herten, W. Germany	30,000	1989
INDAVER	Antwerp, Belgium	50,000	1990
Cleanaway	Brentwood, U.K.	40,000	1990
GSB	Ebenhsusen, W. Germany	100,000	1991
HIM	Wiesbaden, W. Germany	45,000	1991

[a]Value × 1.102 = tons/y.   Source: Short [16].

A. Overview of Permanent Rotary Kiln Incinerators

Typically, general purpose rotary kiln hazardous waste
incinerators installations serve customer corporations that
fit one of two designations: (1) corporations that generate
insufficient quantities of waste to warrant construction of
an on-site incinerator; or (2) companies that choose to focus
their corporate energies and financial resources in their core
lines of business while leaving waste management and disposal
to those firms which have hazardous waste management as a
principal business activity. Centralized general purpose
hazardous waste incinerators are likely to exist within a
larger waste treatment context, as is shown in Fig. 1. Other
processing blocks may include inorganic liquids treatment,
solidification/stabilization, and landfilling of the final
solids generated. Like municipal waste incineration, solid

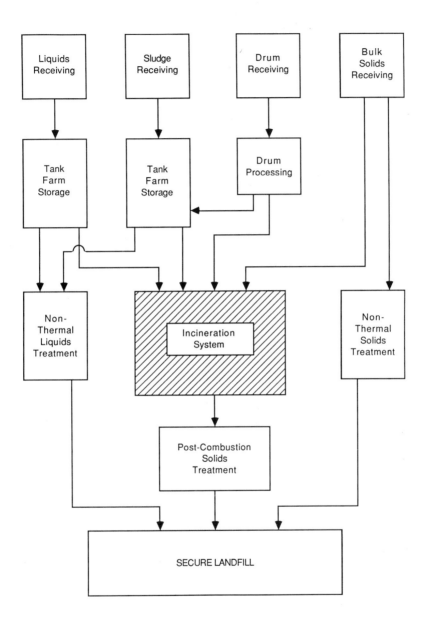

FIGURE 1. Generalized flowsheet of a hazardous waste treatment complex.

hazardous waste incineration always must be supported by a
landfill system.

The incinerator is typically at the heart of such a
treatment complex. It provides and receives necessary support
from the other elements of the treatment facility. The incineration system highlighted in Fig. 1 is, in reality, a very
complex installation as shown in Fig. 2. It involves numerous
unit operations for waste receiving and materials handling,
the incinerator itself, a heat rejection system (waste heat
boiler or quench), an air quality control system, a complex
and detailed instrumentation and controls system, plant utilities, a bottom ash handling and disposal system, administra-
tive services, and other general facilities (i.e., maintenance shops, etc.). The essential ingredient, however, is the
thermal destruction reactor system itself. Rotary kilns are
the dominant reactor system employed (Table II).

B. Rotary Kiln Technology for Hazardous Waste Destruction

A rotary kiln is simply a rotating cylinder which is
typically refractory lined. The cylinder is tilted slightly
(i.e., 3°). Material is fed into the upper end. A heat
source is applied to the material, usually by combusting

TABLE II. Numbers and Capacities of Various Hazardous
Waste Incineration Systems in the U.S.

Technology	Number of units		Ave. design capacity (MMBtu/h)[a]	Reported utilization (%)
	Reported	Projected		
Reported for solids				
Rotary kiln	42	45	61.4	77
Hearth	32	34	22.8	62
Fluidized bed	14	15	19.3	n/a
Other				
Liquid injec- tion	95	101	28.3	55
Fume	25	26	33.1	94

[a]Value × 1.055 = GJ/h.   Source: Oppelt [13].

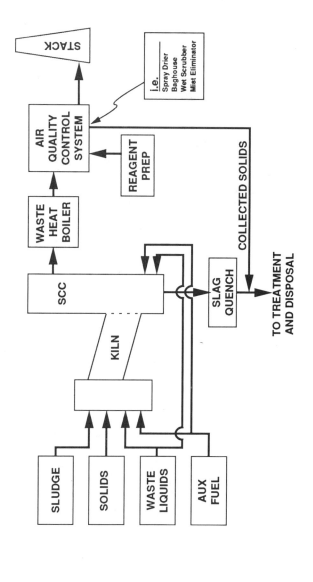

FIGURE 2. *Simplified flowsheet of incineration train in a hazardous waste treatment complex.*

liquid or gaseous fuel within the kiln. The material moves
by gravity through the cylinder and then is discharged from
the lower end. When rotary kilns are used for hazardous
waste incineration, the gaseous products from the kiln are
ducted to a secondary combustion chamber (SCC) for subsequent
destruction (Fig. 2). Rotary kilns are a common thermal
reactor system, with previous uses including (not exhaustive)
cement manufacture, lime calcining for the pulp and paper
industry, municipal waste combustion, mineral ore roasting,
and coal gasification.

The typical hazardous waste kiln is a cylinder which can
have an outside diameter (OD) of 11-12 ft (3.4-3.7 m), and
inside diameter of 8-9 ft (2.5-2.8 m) and a length of 33-38
ft (10.2-11.7 m). Typically, it can have a volume of about
2,000 ft^3 (61 m^3). If 5% of the volume is occupied by solids,
a typical charge can be 100 ft^3 (3.1 m^3) of waste, or 5 tons
(4.5 tonnes) of material. However, the actual load is determined by thermal input to the kiln rather than mass input.
Further, the heat source is a hazardous waste; or an oil or
natural gas flame directly axially down the kiln. The typical
hazardous waste secondary combustor is sized sufficiently to
achieve 2 second residence time for combustion gases at 2200°F
(1200°C).

1. Case Studies in Rotary Kiln Incineration

Numerous rotary kiln hazardous waste incinerators have
been installed and are operating. Among the first was the
Kommunekemi facility in Nyborg, Denmark. Others of note
include the HIM facility in West Germany and the SAKAB facility in Nortorrp, Sweden, described below.

a. *SAKAB AB.* The Svensk Avfallskonvertering AB (SAKAB)
facility handles all of the hazardous waste from Sweden,
including drummed waste, liquids, sludges, and bulk solids.
It is centered around a Von Roll slagging rotary kiln capable
of processing 46,000 tons/yr of hazardous waste. The facility
includes receiving stations, drum processing, tank farms, the
incinerator and post combustion control system, and associated laboratories and administration buildings. The SAKAB
facility (Figs. 3 and 4) came on line in 1984.

The SAKAB facility accepts solid, liquid, and sludge forms
of hazardous waste from all of Sweden. Waste comes in drums.
It may also come as bulk liquids, bulk sludges, or bulk
solids. All wastes are carefully manifested and labelled to
minimize any potential for accidents. Bulk liquids and
sludges are sent directly to the tank farm. Bulk solids and
sludges are sent directly to the tank farm. Bulk solids are

FIGURE 3. Overview of the SAKAB facility in Nortorrp,
Sweden. (Photo courtesy of Environmental Elements Co.)

deposited in the incinerator pit. Drums are opened. Some
are emptied partially or completely depending upon the waste
components. Of interest is the fact that SAKAB uses an automated drum handling system for conveying, opening, and managing the drums (Fig. 5). Liquids and sludges removed from the
drums are sent to the tank farm along with the bulk liquids
and sludges. In the tank farm, extensive blending is performed in order to provide some control over critical parameters of waste composition: heat content or calorific value,
halogen content, nitrogen content, moisture, and inerts.
Extensive chemical characterization and testing is performed
on the waste received to minimize the potential for mixing
incompatible wastes and otherwise causing upset conditions.

For incineration, all wastes are incinerated by going
through the front wall of the Von Roll slagging rotary kiln
at SAKAB. These wastes include not only the solids, sludges,
and liquids; but also the drums carrying much of the waste
to the Nortorrp, Sweden facility. In the kiln, wastes are
burned at elevated temperatures of about 2550°F (1400°C)

FIGURE 4. The SAKAB facility in Nortorrp, Sweden, with
the tank farm and stack in foreground. (Photo courtesy
of Environmental Elements Co.)

FIGURE 5. Drum handling at SAKAB. (Photo courtesy of
Environmental Elements Co.)

FIGURE 6.   Combustion in the slagging rotary kiln at
SAKAB.   Operating temperatures are about 2550°F
(1440°C).   Note the slag in the lower left of the
picture.   (Photo courtesy of Environmental Elements
Co.)

(Fig. 6). This temperature is the hottest generally reported
among centralized facilities [16]. Solids are retained in
the slagging rotary kiln for 1-2 hours while gaseous products
of combustion have a kiln residence time exceeding 2 seconds.
Gaseous products of combustion are then ducted to a secondary
combustion chamber, a waste heat boiler, and then the air
quality control system. The complete SAKAB facility includes
solids and liquids treatment, and landfilling capability (see

Fig. 1 for flowsheet of typical system). The incinerator,
however, is the heart of that system.

The SAKAB facility illustrates the advantages and disadvantages of rotary kiln central incinerators. It has the
ability to destroy virtually any waste fed to it. It has a
high capacity and high reliability. At the same time,
because of its high operating temperature, it has significant
operating costs, particularly when one considers that the
refractory in the unit is replaced every 3,000 hours [16].

   b. *Other Regional Hazardous Waste Facilities.* The SAKAB
plant has benefited from the experience of other systems including the Kommunekemi hazardous waste management facility in
Nyborg, Denmark, and the Hessiche Industriemuell GmbH (HIM)
facility in Biebesheim, West Germany.

The Kommunekemi plant was installed in 1975. It includes
drum management facilities, a tank farm, an incineration
complex, and an off-site landfill at Klintholm, fifteen miles
south of the waste treatment facility. The Nyborg plant
includes two rotary kilns, one of Von Roll design and the
second of Widmer and Ernst design, installed in 1982. It also
includes two separate stack gas cleaning systems. The Von
Roll kiln is equipped with an AQCS based upon an electrostatic
precipitator (ESP) while the Wiedmar and Ernst kiln includes
both a spray dryer-absorber and an ESP. Some 86% of all wastes
entering the Kommunekemi facility are incinerated. The plant
is one of the European installations that uses a waste heat
boiler for heat rejection, and generates useful energy:
34,000 lb/hr of steam for use in district heating [6, 20].
The Nyborg facility is operated by Chemcontrol A/S, in part-
nership with the government of Denmark. It has demonstrated
the essential effectiveness of incineration. At the same
time it has demonstrated that waste transportation and tank
storage can be the sources of problems associated with mixing
incompatible wastes, particularly if the manifesting system
for characterizing the waste is not accurately and completely
followed [20]. This 145,000 ton/yr facility has demonstrated
the long-term utility of thermal destruction as a hazardous
waste management technique.

The HIM incineration facility (see Figs. 7 and 8) is part
of a four-facility operation including two inorganic treatment
plants in Kassel and a secure landfill in Mainflingen. The
HIM incineration complex is similar to the SAKAB and Kommunekemi facilities and includes the capability to receive solid,
liquid, and sludge materials in drum and bulk form. It includes a tank farm for waste management and blending. The
incinerators are Von Roll slagging rotary kilns followed by

FIGURE 7. Overview of rotary kilns at the HIM facility
in Biebesheim, Germany. (Photo courtesy of Environmental
Elements Co.)

an extensive AQCS. The kilns are operated at about 2200°F
(1200°C) to ensure complete organics burnout [16].
     There are several regional hazardous waste incinerators
in North America as well. These include the Chemical Waste
Management (SCA) operation in Chicago, based upon a Combustion
Engineering designed slagging rotary kiln; the Rollins facilities in Deer Park, Texas, Baton Rouge, Louisiana, and Bridgeport, New Jersey, the TSCA incinerator built by a joint
venture of National Electric Company and Westinghouse using
Deutsche-Babcock technology in Coffeyville, Kansas; the Swan
Hills complex in Alberta, Canada; and other regional plants.
Of particular note is the use of drum shredding in the
Chicago operation of Waste Management. This may offer some
materials handling advantages where explosion hazards can be

FIGURE 8.   Bulk sludge receiving at the HIM facility.
(Photo courtesy of Environmental Elements Co.)

successfully managed. Drum shredding is not a widely practical feed preparation system, at present.

The Swan Hills facility in Alberta, Canada, provides insights into a small regional facility. The incineration
system is part of a multiprocess complex including physical/
chemical treatment, stabilization, landfilling, and deep well
injection. The incineration system itself is quite small.
It uses two small batch Von Roll rocking kilns, each having
a capacity of 8.75 million Btu/hr, or 5,500 tons (5,000 tonnes)
of hazardous waste per year. The kilns are fed specific
wastes. They rock (rather than rotate) approximately 45°
from the centerline in each direction. The kilns operate at
temperatures ranging from 1100°F to 2350° (610–1300°C) and
the secondary combustion chamber operates at 2200°F (1210°C).

Ash is discharged from the primary reactor by rotating the
kiln 120-180°. The ash and slag flows into a water quench at
that time. The Swan Lake facility design has the ability to
completely control combustion, and to minimize "puffs" and
"burps"--sudden expansion of combustion gases passing through
the entire operation. This plant further demonstrates the
ability to construct small permanent installations for hazardous waste combustion [2, 14].

In summary, then, there is a significant body of experience that has been established demonstrating the utility of
permanent rotary kiln incinerator installations. These installations employ a wide range of technological innovations
in the management and destruction of hazardous waste.

2. Rotary Kiln Process Alternatives

There are several types of rotary kilns and associated
afterburners used in hazardous waste incineration as highlighted by the case studies. Kiln types include cocurrent
and countercurrent flow rotary kilns, rocking kilns, slagging
rotary kilns, and ashing rotary kilns. Cocurrent and countercurrent flow kilns are distinguished by the placement of the
heat source or flame, as shown in Fig. 9. Rocking kilns are
distinguished from conventional kilns in that they are batch
fed rather than continuously or semi-continuously fed. Slagging and ashing kilns are distinguished by temperatures within
the reactor. Slagging kilns typically operate at about 2200-
2600°F (1203-1430°C) while ashing kilns operate in lower
temperature regions of 1200-1800°F (650-980°C). Kilns
designed for slagging operation can have the capability to
operate in either mode. The consequences of operating in the
slagging mode include more complex kiln front walls (Fig. 10),
more complex interfaces between kiln and secondary combustion
chamber (Fig. 11), and more stringent refractory requirements.
Further, maintenance (including refractory replacement) is
more expensive when slagging operation is used. For example,
slagging kilns may require refractory replacement on a 3,000-
5,000 hour schedule [16] depending upon operating temperature
and feedstock characteristics.

All kiln systems include secondary combustion chambers or
afterburners. Afterburners are characterized largely by
orientation, with designs including horizontal afterburners,
vertical upflow afterburners (Fig. 11), and vertical downflow
afterburners (Fig. 12). The primary selection points among
rotary kiln design alternatives are flow directions and temperature regimes. Countercurrent flow kilns maximize heat and
mass transfer, having a distinct "cold end" where solids are
introduced and gaseous products are discharged, and a

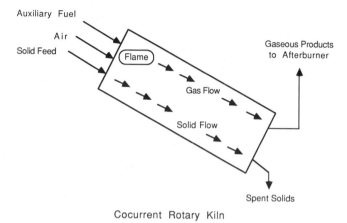

Cocurrent Rotary Kiln

Countercurrent Rotary Kiln

*FIGURE 9. Gas and feed flows in countercurrent and
cocurrent rotary kilns.*

"hot end" where combustion occurs and where heated solids are
discharged. The conceptual temperature profile of such kilns
is shown in Fig. 13. Countercurrent kilns, therefore, maximize energy efficiency. At the same time, however, they
require complex materials feed and discharge systems.

FIGURE 10. Inside view of the front wall of a Von Roll slagging rotary kiln. Note the solids feed chute, the sludge lance, the liquids and auxiliary fuel burners, and the combustion air inlet. (Photo courtesy of Environmental Elements Co.)

Further, the presence of a flame at the solids discharge end along with the need for ducting the gaseous products of combustion from the solids feed end to a secondary combustion chamber significantly complicates the materials handling aspects of kiln design. Finally, there is considerable difficulty in manipulating the combustion mechanism and controlling the incineration residence time for solids in a countercurrent kiln, particularly if significant quantities of solids are entrained in the gaseous products of combustion.

FIGURE 11. The rotary kiln/secondary combustion chamber
transition at HIM. Note the volume of the secondary
combustion chamber. Also note the secondary air parts
above the kiln discharge. (Photo courtesy of Environmental Elements Co.)

Consequently, countercurrent flow kilns are more commonly used
only in minerals production and refining (i.e., cement manufacture, lightweight expanded aggregate processing, limestone
calcining) than hazardous waste incineration.

Cocurrent kilns, as shown in Fig. 14, introduce waste and
any auxiliary fuel at the same end as discussed previously.

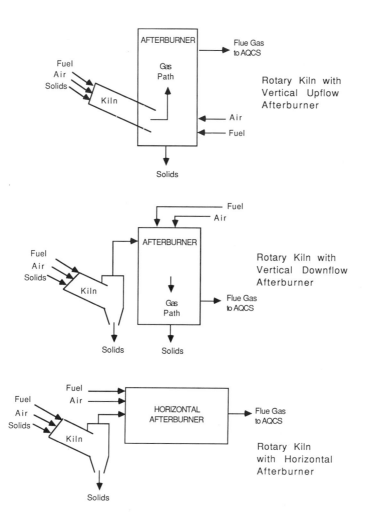

FIGURE 12. Secondary combustion chamber or afterburner configurations.

There is no distinct "cold end" and "hot end" in cocurrent kiln operation. The entire unit is maintained in the reaction temperature region. Cocurrent kilns, which predominate in hazardous waste incineration permit careful manipulation of the combustion mechanism discussed in Chapter V.

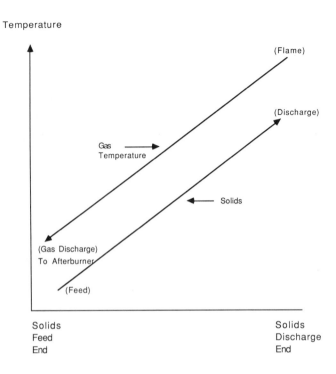

FIGURE 13. Conceptual temperature profile of a counter-current kiln.

The selection of temperature regime--slagging or nonslag-ging operation--is a function of the types of material being processed and the regulatory environment. When the feedstocks being processed have a high calorific value (higher heating value greater than about 5,000 Btu/lb or 2.8 kcal/g as received) along with moderate moisture and halogen contents, and can sustain high temperature combustion, or when the materials being processed include drums of waste, then slag-ging kilns are favored. Slagging kilns can achieve higher throughputs of significant calorific value wastes than ashing kilns, and they can treat both the drums and the drummed wastes directly. When the feedstock is largely low calorific value wastes such as contaminated dirt, and particularly when the operation of the kiln will require a constant use of

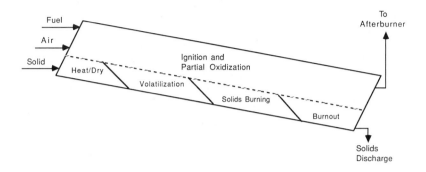

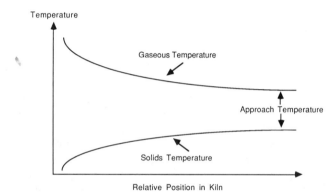

FIGURE 14. Stages of combustion and conceptual temperature profile of a cocurrent rotary kiln.

substantial auxiliary fuel for temperature maintenance, ashing
kilns are more economical to operate.

C.   Combustion in Rotary Kiln Systems

The three Ts of combustion--time, temperature, and turbu-
lence--are carefully accommodated by kiln system design and
operation. Further, the stages of combustion are distinct
within the kiln as shown previously in Fig. 14. Stages of
combustion along with all other combustion process considerations are generally optimized when the gaseous velocity
within the kiln is held to rates of about 15 to 20 ft/sec.
Heat release levels are generally held to about 25,000-
40,000 Btu/cu ft/hr [8]. In manipulating combustion with a

rotary kiln, particular emphasis is placed on the following
variables: (1) kiln dimensions and length/diameter (L/D)
ratio; (2) kiln cross sectional loading; (3) solids residence
time in the kiln, including the presence or absence of flights
or augers within the kiln for materials manipulation; and
(4) the use of excess air within the kiln. All of these variables relate to the three Ts of combustion, to the achieving
of 99.99% DRE, and to the minimization of airborne emissions
including particulates, acid gases, and products of incomplete combustion (PICs).

1. Temperature Manipulation

Temperature manipulation is largely a function of refractory and system limitations. Higher temperatures permit more
kiln throughput. For wastes with sufficient calorific value
to sustain combustion (i.e., pentachlorophenol, napthalene),
higher temperatures permit increased volumes of material per
unit time within the limits of heat release and gaseous velocity. Higher temperatures become a problem, however, when
largely inert materials are being fed to the kiln. Such materials include the contaminated soils and sludges where POHC
concentrations may be in the 100-1,000 ppmv range.

Temperature manipulation also impacts the fate of heavy
metals, with higher temperatures volatilizing increasing
quantities and percentages of such materials as lead, zinc,
cadmium, chromium, etc. Temperature is a function of the
inputs to the kiln. These inputs include the waste feed, any
supplementary fuel, atomizing steam if used, and excess air.
The waste feed and supplementary fuel can be combined by calculation into an "effective feed" stream. It can then be
shown that the gaseous temperature of the kiln can be estimated by the following equation:

$$T_{kg} = 234.8 + .097(C) + 3,518(\phi) \qquad\qquad (6-3)$$

Where T_{kg} is kiln gas temperature in °F, C is the higher
heating value or calorific value of the effective feed and
ϕ is equivalence ratio. Alternatively, equation 6-3 can be
written as follows:

$$TW_{kg} = 3,058 + .095(C) - 9.68 \text{ (EA percent)} \qquad (6-4)$$

Where EA percent is percentage of excess air. These equations were derived from heat balances about kilns incinerating soils contaminated with 2-chlorophenol at levels of 30%
to 67%. The excess air levels associated only with the kiln
ranged from 50% to 150% (equivalence ratios of 0.67-0.4).

These equations provide useful temperature approximations,
recognizing that actual temperatures will vary moderately
about the values calculated as a function of halogen and
moisture content in the feed, the use of atomizing steam, and
specific kiln performance.

Rotary kilns can achieve temperature control first by
manipulation of the calorific value and moisture content of
the effective feed. When the waste carries sufficient calorific value to support combustion at required temperatures,
effective feed manipulation by blending becomes essential.
Very high calorific values can produce excessive temperature
even in slagging reactors. When the material being incinerated cannot sustain combustion without the use of supplementary fuel, calorific value is manipulated by the rate at
which the supplementary fuel (i.e., No. 2 distillate oil or
natural gas) is fed to the primary reactor (relative to the
waste feed). It is also useful to note that the excess air
within the kiln is another effective moderator of temperature.
Each percentage increase in excess air decreases kiln gas
temperature by about 10°F, largely due to the presence of
inert nitrogen in the gas stream.

Temperature of the solids is manipulated by solids residence time. Given a specific gas temperature, solids temperature at the discharge end is controlled by utilization of the
approach temperature phenomena (see Fig. 14). Longer solids
residence times will decrease the approach temperature while
shorter solids residence times will increase the approach
temperature. In slagging kilns, with solids residence times
of 60 to 120 min., there may be no real approach temperature.
In ashing kilns, with solids residence times of 30 to 60 min.,
approach temperatures of 300-500°F are common.

It is necessary to control temperatures by combined manipulation of the calorific value, the excess air level, and the
solids residence time particularly when the waste being incinerated contains less than 1,000 Btu/lb and requires the addition of auxiliary fuel. Under such conditions the kiln is
typically operating in the ashing (nonslagging) mode, is
using supplementary fuel, and has an afterburner operating at
temperatures that exceed the exit gas temperatures of the
kiln. Because of the differential in heat transfer effectiveness between solids and gases, it is more fuel efficient
to raise temperature in the kiln than in the afterburner [19].
Consequently, it may be desirable to "push" the kiln and
minimize the temperature differential between kiln and afterburner.

Afterburner temperature manipulation is achieved by controlling the temperature and composition of the exit gas from
the kiln, and by the use of either supplementary fuel or

light liquid wastes (i.e., acetone, methyl-ethyl-ketone).
Alternatively, temperatures are reduced by the introduction
of steam, water, and excess air; or by heat removal in high
temperature heat recovery chambers. Afterburner temperatures
must be maintained in excess of 2250°F (1230°C) for 2 seconds
for TSCA wastes and are maintained generally in the region
of 1600-2800°F (870-1540°C) for RCRA wastes [8]. Exit gas
temperatures for afterburners or secondary combustion temperatures, however, may be reduced to the range of 1350-1400°F
(730-760°C) in central station incinerators with large afterburners having a residence time exceeding 2 seconds, and in
installations with waste heat boilers, in order to optimize
downstream equipment.

2. Residence Time Manipulation

Given the kiln velocity limitation of 15-20 ft sec.,
temperature and residence time are particularly important
variables for manipulation. The basic kiln dimensions of
effective length and inside diameter establish the volume
available for gaseous and solid products of combustion. The
cross sectional kiln loading, shown in Fig. 15 and typically
expressed as a percentage, establishes the volume of the kiln
available for gaseous products of combustion.

Gaseous residence time is, effectively, a function of
kiln velocity; and it can be calculated as follows:

$$T_{r,g} + V_g \times L_k \tag{6-5}$$

Where $T_{r,g}$ is gaseous residence time in the kiln, measured in
seconds, V is velocity of the kiln gases, measured in ft/sec,
and L_k is effective kiln length. V_g, in turn, is calculated
as follows:

$$V_g = (KG_{1b-m,s} \times 359 \times [(T_{kg} + 460)/492])/$$

$$[\pi R_k^2 \times (1-Lo/1000)] \tag{6-6}$$

Where $KG_{1b-m,s}$ is pound-moles of kiln gas generated by combustion per second, T_{kg} is temperature of the kiln gas on °F, R_k
is inside radius of the kiln, and Lo is kiln loading expressed
on a percentage basis. $KG_{1b-m,s}$ is calculated from the
hourly heat balance of the reactor as follows:

$$KG_{1b-m,s} = KG_{1bm,h}/3,600 \tag{6-7}$$

Given the principles governing velocity and residence time in
the kiln, it is useful to evaluate influences on residence

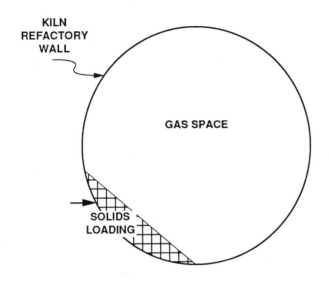

FIGURE 15. Schematic of kiln loading. Note that the loading level shown is 7 1/2%.

time and velocity. The dominant influence is the total volume of the rotary kiln itself. The second influence is that of kiln loading, however, kiln loading and, coincidentally, feed rate, has a relatively minor impact because kiln loading for ashing reactors typically ranges from 7.5% to 15%, and for slagging reactors loading levels are typically in the 4% to 6% region. Loading becomes important here, because it determines the exposure of solids to the heat sources: the flame and the hot kiln walls. Higher loading results in less exposure. Since the influence of loading is measured by the void space in the kiln, the range available for ashing kilns is typically 85% to 92.5% and the range available for slagging kilns is 94–96%.

The influence of temperature, when moderated by excess air, is virtually insignificant due to the countervailing

forces exhibited by these two variables. Increasing excess
air increases the lb-moles of gas passing through the reactor
per unit time (i.e., lb.m/sec); however, the excess air decreases the temperature and the consequent volume of those
gases (see equation 6-6). Illustrating this phenomenon are
calculations concerning a kiln 12 ft in diameter and 40 ft
long, fed soils contaminated with chorobenzene having an
effective feed calorific value of 1,550 Btu/lb. Given a
feed rate of 25 million Btu/hr (a loading of 10.6%), the velocity can be calculated by the following expression:

$$V_{kg} = 14.20 + 0.20 \ (EA \ percent) - 0.006T_{kg} \qquad (6-8)$$

The increased excess air and the decreased temperature are
offsetting factors.

Temperature, velocity, and residence time can be altered,
as can temperature, by using oxygen enrichment of the air
[1, 9]. Use of an oxidant with 10-40% pure oxygen and 90-60%
air can increase temperature, decrease auxiliary fuel requirements, and improve throughput. The use of oxygen enrichment
decreases the gaseous volume in the kiln, and the gas velocity,
by reducing the nitrogen load in the oxidant and products of
combustion. The driving factor here is lb-m N_2/lb-m O_2 in the
oxidant. Reduction in gaseous volume by decreasing the lb-m
N_2/lb-m O_2 increases kiln throughput by maintaining the 15-
20 ft/sec gas velocity in the kiln while firing more waste
with less volume of oxidant. The equation governing this
phenomenon is as follows:

$$lb\text{-}m \ N_2/lb\text{-}m \ O_2 = 3.76 - .038 \ (\% \ O_2E) \qquad (6-9)$$

Where % O_2E is percent pure oxygen enrichment of the oxidant.
Due to kiln in-leakage, 40% oxygen enrichment appears to be
a practical upper limit [9]. Residence time and velocity in
the afterburner are similarly manipulated largely by the
volume of the reactor and, if desired, the use of oxygen
enrichment. Typically, the regulations require 2 seconds
residence time of 2250°F (1230°C) for TSCA wastes, and sufficient residence time to ensure 99.99% DRE for RCRA wastes.
Further, ample residence time and temperature in the secondary reactor are required for PIC minimization.

Solids residence time in the kiln is manipulated independently of gaseous residence time. Solids residence time can be
calculated by the following equation, developed by the U.S.
Bureau of Mines:

$$R_s = (1.77 * \Theta \times L \times F)/(S \times D \times N) \qquad (6-10)$$

Where R_s is residence time of solids, expressed in minutes
Θ is the angle of repose of the solid hazardous waste (typically taken at 35° for soils), L is kiln length, S is the slope
of the kiln (degrees), D is the kiln diameter, N is the kiln
rotational speed in RPM, and F is a factor taken to adjust for
the presence of lifters, flights, or dams in the kiln. F is
taken as 1.0 for smooth kilns and may be reduced to about 0.5
with the presence of lifters. The F factor raises the particularly significant issue of modifying the kiln internals to
increase residence time, radiative heat transfer to the
materials being incinerated, and thereby effectively decreasing the approach temperature. Kiln internals are commonly
used in rotary kiln design, particularly in ashing kilns.

Common solids residence times range from 60 to 120 minutes
for centralized slagging rotary kilns. Two-hour residence
times for solids appear to be the most common [8]. Minimization of solids residence time, however, is particularly useful
for maximizing the cost-effectiveness of rotary kiln incinerators.

Temperature and time calculations lead to the typical heat
and material balances associated with rotary kiln incineration
systems. Figure 16 is a heat balance about a hypothetical
slagging rotary kiln burning a mixture of hazardous wastes, some
contained in drums. Wastes include liquids, sludges, and
bulk solids. Figure 17 is a heat balance about a "dirt burning" ashing kiln. These heat and material balances illustrate
the operation of centralized rotary kiln hazardous waste
incinerators.

3. Other Considerations

Final considerations of significance are kiln pressures,
length/diameter (L/D) ratio of the kiln itself, and kiln
refractory. Kiln pressures are typically maintained at
−0.5 to −2.0 in w.c. [8]. Slightly negative pressures mean
that the system is largely driven by the induced draft fan.
These negative pressures have important safety implications,
however, due to the potential for gases escaping through kiln
seals unless leakage consists of infiltrating the kiln rather
than products of combustion escaping. L/D ratios are significant in terms of the residence time in the reactor of both
solids and gases, capital requirements, and also airborne
emission formation. It is significant to note that lower L/D
ratios minimize the formation of particulates, largely by
reducing the velocity of the gases through the kiln and hence
the entrainment. Low L/D ratios of kilns with lifters or
flights, however, must be adjusted for the F factor, as such
flights necessarily pass the solids through the gas stream.

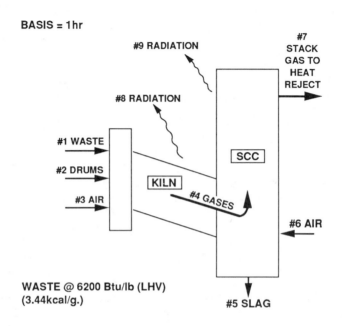

BASIS = 1 hr

#9 RADIATION

#7 STACK GAS TO HEAT REJECT

#8 RADIATION

SCC

#1 WASTE

#2 DRUMS

KILN

#3 AIR

#4 GASES

#6 AIR

WASTE @ 6200 Btu/lb (LHV)
(3.44kcal/g.)

#5 SLAG

STREAM	MASS (ton)	TEMP (°F)	HEAT FLOW (BTU x 10^6)
1. WASTE	6	77	74.4
2 . DRUMS	0.2	77	-0-
3 . AIR	43.5	77	-0-
4. COMB. GAS	47.6	2550	68.4
5. SLAG	2.1	2550	2.3
6. AIR	14.5	77	-0-
7. STACK GAS	62.1	2000	65.5
8. RADIATION	NA	NA	3.7
9. RADIATION	NA	NA	1.4

FIGURE 16. Representative calculated heat and mass balance about a slagging rotary kiln.

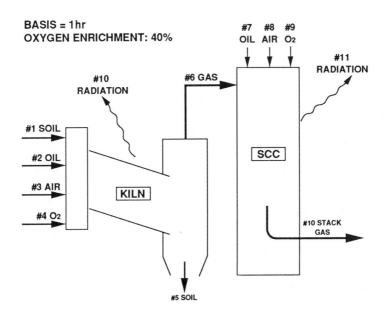

STREAM	MASS (ton)	TEMP (°F)	HEAT FLOW (BTU x 10^6)
1. SOIL	16	77	-8.4
2. #2 OIL	1.1	77	40
3. AIR	13.8	77	0
4. OXYGEN	2.1	77	0
5. CLEAN SOIL	12.2	1400	6.4
6. STACK GAS	20.9	1800	23.2
7. #2 OIL	0.8	77	29.2
8. AIR	13.0	77	0
9. OXYGEN	2.0	77	0
10. STACK GAS	36.7	2250	1.2
11. RADIATION	NA	NA	2.0
12. RADIATION	NA	NA	1.0

FIGURE 17. Representative calculated heat and mass balance about an ashing rotary kiln being operated as a "dirt burner," and using oxygen enrichment.

Typical L/D ratios in current designs are in the range of 3.4-
4.0, although ratios as high as 8.0 have been proposed.

Kiln refractory is determined by the operating requirements of the unit, with particular attention to feed stream
condition, chemical composition, and temperature of operation.
As drums are included in the feed stream, abrasion resistance
must be increased. Significant concentrations of certain compounds (i.e., sodium, fluorides) place increasing demands upon
the refractory. Higher temperatures also impact refractory
selection and refractory life as shown, for example, at the
SAKAB facility where refractory replacement is every 3,000
hours. Typical installations use insulating firebrick covered
by 6-10 inches of high temperature, highly erosion-resistant
firebrock where appropriate [10]. Slagging rotary kilns may
"freeze" a layer of slag onto the refractory in order to protect it and extend its life.

4. *Data Requirements for Incinerator Design
    and Operation*

The care going into the design and operation of a centralized hazardous waste incineration facility leads to the
requirements for incinerability data used to support regime
determination. Incinerability data for individual compounds,
and classes of compounds, permits regime determination for
economic and environmental optimization of the facility. Such
data can be applied most precisely in industrial facilities
accepting a limited range of feedstocks. At the same time,
such data can be useful in establishing operating parameters
for centralized facilities.

Virtually all vendors maintain pilot facilities for determining the incineration characteristics of anticipated feed
streams. Such data are developed in rotary kilns where the
scale-up parameters are well documented. Additionally, firms
such as EER Corporation (see Fig. 18), Acurex, and others;
and universities and academic organizations such as the
University of Dayton Research Institute (UDRI), University of
Utah, and Louisiana State University; provide extensive facilities for the testing of harzardous waste incineration.
Results of such testing operations provide essential insights
improving the state-of-the-art in high temperature incineration.

D. Availability and Cost of Rotary Kiln Systems

Rotary kiln technology is well established, as noted previously. Consequently, kiln systems are readily available to
prospective incineration developers. Table III is a partial

FIGURE 18. *The EER rotary kiln simulator and test facility.*

list of rotary kiln system vendors offering either slagging, ashing, or both types of systems.

Capital costs, particularly for centralized rotary kiln systems, are highly project specific and depend upon the following factors: capacity, types of feedstocks being fed, regime (slagging vs. ashing), L/D ratio, type of solids discharge system, type and capacity of afterburner, type of auxiliary fuel used, and regulatory climate. Total systems can cost in the range of 30-70 cents/Btu of capacity, or $750-$1,250/annual ton of capacity depending upon factors cited above.

Operating costs for rotary kiln installations are driven largely by labor requirements, and by maintenance (i.e., refractory replacement). With taxes, insurances, and profits included, these parameters lead to typical incineration costs at fixed facilities of $0.15-$0.33/lb, or $300-$650/ton [15]. Such costs are based upon chemical composition of the waste plus the volume of waste being incinerated. Specific, difficult wastes may be more expensive to incinerate. If a permanent installation is constructed by a manufacturer for its own hazardous wastes, costs may range from $0.20 to $1.00/lb, or $400-$2,000/ton depending upon the nature and volume of the waste being incinerated. On-site incinerators may

TABLE III.   Partial List of Rotary Kiln Vendors for
Permanent Rotary Kiln Installations

Vendor name	Vendor location
Allis-Chalmers	Milwaukee, Wisconsin
Combustion Engineering Co. Raymond Div.	Chicago, Illinois
Deutche-Babcock (Ford, Bacon, and Davis)	Salt Lake City, Utah
Environmental Elements Co.	Baltimore, Maryland
Fuller Power Company	Bethlehem, Pennsylvania
Industronics, Inc.	S. Windsor, Connecticut
International Waste Energy Systems	St. Louis, Missouri
John Zink Company	Tulsa, Oklahoma
Von Roll Otd.	Zurich, Switzerland
Von Roll, U.S.A.	New Jersey

typically have total costs of less than $0.30-$0.50/lb, and
may approach the cost of centralized facilities, when annual
volumes of hazardous wastes to be incinerated exceed 10,000-
20,000 tons per year [15].

Rotary kilns have become the technology of choice for general purpose incineration systems. They are well proven
through many installations. They can handle the widest variety
of waste, and can be built to large capacities. They are
readily managed with respect to manipulating the combustion
mechanism and they can be purchased from a significant number
of vendors. They are, however, not inexpensive.

III.   FLUIDIZED BED HAZARDOUS WASTE INCINERATION SYSTEMS

Fluidized bed incineration systems offer an emerging
alternative to rotary kilns. Fluidized beds are vessels containing an inventory of inert material such as sand or sized
limestone particules. Combustion air is introduced from the
windbox under the bed at sufficient velocities to "fluidize"
the bed media. Fluidization typically occurs at a superficial velocity of about 3-10 ft/sec. Higher velocities may

be used depending upon design. Fuel or combustible material
is introduced either directly into the bed (in-bed feeding)
by augers and screw feeders or is flung over the bed (overbed
feeding) by stokers. The solid fuel is subjected to intimate
contact with the bed media, causing immediate heat transfer
from the bed particles to the fresh fuel. Ignition and combustion occur. At the same time the bed media "scrub" the
fuel particles, constantly exposing fresh fuel surface by the
abrasion process. In this way very high combustion efficiencies occur. Typical fluidized bed sketches have been shown
previously in Chapter IV.

A. Technology Overview

     Fluidized bed technology has existed for well over 50
years, with applications including coal gasification (the
Winkler gasifier was introduced around 1930), petroleum
refining, solid particle drying, and the combustion of solid
fuels. In recent years fluidized bed combustors have been
constructed to burn a wide variety of materials including
coal, heavy oil, municipal waste (see Chapter IV), coal
washery waste, culm, and other high ash materials. These
reactors have been installed at capacities ranging from 50
million Btu/hr (53 GJ/h) of heat input to 1.1 billion Btu/hr
(1.2 TJ/h). The largest unit constructed to date is the
Colorado-Ute 110 Megawatt (MWe) coal-fired electricity generating station. One of the early fluidized beds applied to
hazardous type wastes was the Ahlstrom-Pyropower design installed in 1980 at Kemira Oy, Finland, for the incineration
of zinciferous sludge.

1. *Alternative Fluidized Bed Systems*

     Several basic types of fluidized bed combustors exist
including: (1) bubbling beds; (2) circulating fluidized beds;
(3) multistage beds; (4) spouting beds; and (5) pressurized
fluidized beds. Of these, the bubbling beds and circulating
beds as discussed in Chapter IV are most appropriate and most
commonly applied to solid hazardous wastes.
     Fluidized bed alternatives are distinguished largely by
the superficial velocity of the gas stream, and the consequent handling of solids. Bubbling beds have low velocities
(i.e., 3-10 ft/sec), and maintain the bed in a dense fluid
phase. Circulating fluidized beds have much higher in-bed
velocities (i.e., 15-30 ft/sec or 4.5-9 m/sec) than bubbling
beds. They are designed to blow much of the bed into the hot
cyclone, capture the solids in the large cyclone, and then
recycle the solids to the primary reactor. Circulating

fluidized bed media particles may be recycled from 15 to 30
times through the reactor; and fresh bed material is introduced at a very modest rate.

2. Application of Fluidized Beds to Waste Incineration

Fluidized bed incinerators have been applied to a wide
variety of unusual and/or hazardous wastes, as well as municipal wastes (see Chapter IV). For example, Keeler-Dorr Oliver
reports installations burning spent cutting oils, paint
wastes, steel mill wastes, biological sludge, oil refinery tank
bottoms, spent caustic, paper mill sludge, and a variety of
chemical sludges. Waste burning installations have ranged in
size from 450 lb/hr (205 kg/hr) to 3.5 tons/hr (3.2 tonnes/hr).
Ogden Environmental (formerly GA Technologies) has applied
the Pyropower circulating fluidized bed design to a host of
materials including PCB contaminated soils. Fluidized bed
incineration is applicable to a wide variety of hazardous
wastes as well.

It is useful to note that fluidized beds are particularly
amenable to waste feeds with potential for high nitrogen content. Fluidized beds typically operate at temperatures of
1550-1800°F (840-980°C), where oxides of nitrogen cannot be
formed in appreciable quantities from thermal NO_x sources.
Further, fluidized beds have been designed incorporating
ammonia injection within the fluidized bed for control of any
NO_x formed from either thermal or fuel sources. The only
limitation on ammonia injection is the potential for increasing CO emissions associated with the ammonia reduction reactions and the careful interpretation of those data.

B. Combustion of Hazardous Wastes in Fluidized
     Bed Reactors

All fluidized bed hazardous waste incinerators involve a
primary reactor, but avoid the need for an afterburner. These
reactors employ intimate mixing of fuel and bed media at temperatures of 1400-2000°F (760-1093°C) and centering around
1550-1800°F (815-982°C), have combustion gas residence times
of 1-5 sec., and can use excess air levels of 100-150% [10].
Recent fluidized bed experience has shown excess air levels
of 40-70%, however [7]. Such levels of excess air are more
in keeping with efficient combustion practice.

1. Destruction and Removal Efficiency

Fluidized bed combustors, using intimate mixing and high turbulence along with maximized heat transfer from the bed media to the feed, can achieve the required DRE levels on a wide variety of feedstocks. Oppelt [13] reports one on-site fluidized bed operating with 3.6% O_2 in the stack and achieving 99.996% DRE on RCRA wastes. This incinerator produces CO emissions of 67.4 ppm, demonstrating highly successful combustion. Chang et al. [4] report DRE levels up to 99.9999% on soil contaminated with Freon 113, and similar results on soil contaminated with carbon tetrachloride introduced into a pilot scale circulating fluidized bed (see Table IV). The TSCA trial burn results for the Pyropower circulating fluidized bed reactor demonstrated 99.9999% DRE of PCB contamination in soils up to 10,000 ppm [7]. Trial burn results on the Pyropower fluidized bed as applied to RCRA wastes are shown in Table V. Table VI summarizes results of the incineration of PCB soil in the GA incinerator. DRE levels and

TABLE IV. Test Results for Toxic/Hazardous Waste Combustion in GA'S CBC

Chemical name	Chemical formula	Physical form	Destruction and removal efficiency of POHCs[a]	HCl Capture
PCB	$C_{12}H_7Cl_3$	Soil contaminant	>99.9999	>99
Carbon tetrachloride[4]	CCl	Liquid	99.9992	>99
Freon	$C_2Cl_3F_3$	Liquid	99.9995	>99
Malathion	$C_{10}H_{19}O_6PS_2$	Liquid	>99.9999	–
Dichloro-benzene	$C_6H_4Cl_2$	Sludge	99.9999	>99
Aromatic nitrile	$C_8N_2H_4$	Tacky solid	>99.9999	–
Trichloro-ethane	C_2HCl_3	Liquid	99.99999	>99

[a]POHC = principal organic hazardous constituents.
Source: [7].

TABLE V. Circulating Bed Combustor Test Results of Various Wastes

Waste	Cl, F or S (%)	Heating value		HCl, HF or SO$_2$ retention (%)	Volume reduction ratio	NO$_x$ (ppm)	Destruction and removal efficiency or principal organic hazardous constituents (%)
		(Btu/lb)	(kcal/g)				
Depleted uranium slurry	N/A	9,700	5.39	N/A	8:1	100	N/A
Heavy metal waste	2.5	12,300	6.83	94.8	12:1	–	N/A
Oily water sludge	4	3,000	1.67	99.5	15:1	90	99.99+
Chemical plant wastes	1	14,000	7.78	95.8	7:1	70	99.9+
Chlorinated organic sludge	14	14,000	7.78	99.99	N/A	50	99.9999+
Aluminum potlinings	20	8,000	4.44	90	N/A	100	N/A
Thorium/uranium soot	N/A	14,000	7.78	N/A	48.1	100	N/A
Chlorinated coke residue	7	8,500	4.72	99.5	–	120	N/A
PCB-contaminated soil	10,000	0	0	99.9	–	55	>99.9999+

Note: Results were obtained in tests at the GA Technologies, Inc. pilot plant and at a similar unit in Finland.

TABLE VI. CBC Test Conditions and Results: Treatment of
PCB-Soil

	EPA triplicate runs		
Conditions:			
PCB concentration in soil, ppm	11,000	12,000	9,800
Soil feedrate, lb/hr	328	412	324
Combustion temperature, °F	1,805	1,805	1,795
Superficial velocity, ft/sec	18.7	18.7	18.1
Gaseous residence time (sec)	1.18	1.18	1.22
Excess oxygen, percent	7.9	6.8	6.8
Results:			
DRE, percent	99.9999	99.9999	99.9999
Combustion efficiency	99.94	99.95	97.97
PCB in bed ash, ppb	3.5	33	186
PCB in flyash, ppb	66	10	32
NO_x, ppm	26	25	76
CO, ppm	35	28	22
Dioxins and furans	ND[a]	ND[a]	ND[a]
HCl, ppm	57	202[b]	255[b]
Particulates, gr/ft^3	0.09	0.04	0.002

DRE = Destruction removal efficiency of PCB.

[a] None detected.
[b] Intermittent limestone feed.
Source: [7].

other salient combustion parameters associated with applying
the Pyropower incinerator to polychlorinated biphenyl (PCB)
wastes are also shown in Table VI. The major feedstock limitations of CFB technology appear to be drummed wastes and
wastes with the potential to slag during combustion.
    All fluidized beds feature in-bed capture of acid gases
such as SO_2 and HCl by the following reactions:

$$CaCO_3 => CaO + CO_2 \qquad\qquad (6-11)$$

$$SO_2 + CaO + 0.5O_2 => CaSO_4 \qquad\qquad (6-12)$$

$$2HCl + CaO => Ca\ Cl_2 + H_2O \qquad\qquad (6-13)$$

As such, fluidized bed reactors eliminate the need for acid gas scrubber systems in the total process configuration. The reactor reported by Oppelt [13] achieved HCl emissions less than 4 lb/hr. Acid gas control efficiencies as high as 98% have been reported for circulating fluidized bed reactors [18]. Acid gas control efficiencies are governed by concentrations of chlorine and sulfur in the contamination, and by the Ca/S+Cl molar ratios in the bed media.

The consequence of fluidized bed combustion is a simplified process flowsheet marked by the potential for eliminating the secondary combustion chamber and the acid gas control system (Fig. 19). It should be pointed out that this elimination of the afterburner comes with a modest primary reactor penalty. The heat release rate for fluidized bed hazardous waste incinerators is 20,000-35,000 Btu/cu ft-hr (3.6-6.3 GJ/m^3-hr), which is about 80-85% of the heat release rate in rotary kiln systems [8]. Even with this larger primary reactor, on a heat release basis, the fluidized bed system is substantially simpler than the rotary kiln system. This dramatic change in combustion technology can reduce the capital investment in the hazardous waste incinerator by as much as 33% when compared to the rotary kiln technology.

C. Availability of Fluidized Bed Incinerators for Hazardous Waste

There are numerous vendors of fluidized beds that can be applied to hazardous wastes. Conventional "bubbling" beds are offered by Dorr Oliver, Process Combustion, and others as shown in Table VII. Circulating fluidized beds are offered by Ogden Environmental Services (the Pyropower CFB), Combustion Engineering-Lurgi, and Gotaverken. All of these units have been designed and built based upon a significant base of experience with fluidized bed reactors.

Despite the traditions of fluidized bed reactors, and the experience gained to date, the potential for fluidized bed incineration of hazardous waste is only beginning to be explored. Advantages of fluidized bed systems include operation at lower temperatures than rotary kilns and afterburners, and improving the fuel economics when wastes with little or no calorific values are being incinerated. Advantages also include the thermal inertia when the waste varies with an

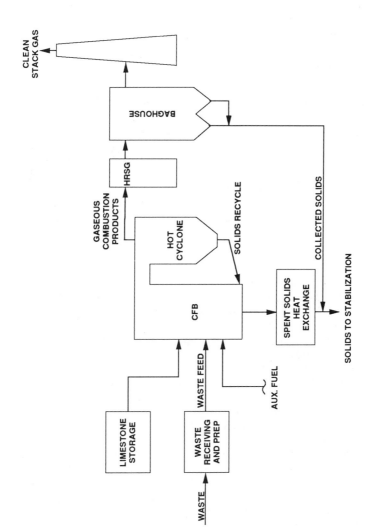

FIGURE 19. Simplified flowsheet about a circulating fluidized bed hazardous waste incinerator.

TABLE VII. Partial List of Fluidized Bed Hazardous
Waste Incinerator Vendors

Vendor	Location
Combustion Engineering, Raymond Division	Chicago, Illinois
Copeland Associates, Inc.	Oak Brook, Illinois
Fuller Company	Bethlehem, Pennsylvania
Keeler-Dorr Oliver	Stamford, Connecticut Williamsport, Pennsylvania
Lurgi Corporation	Belmont, California
Ogden Environmental/Pyropower	San Diego, California
Process Combustion	Pittsburgh, Pennsylvania
Thermal Processes, Inc.	La Grange, Illinois

Sources: [8], [10], and personal files.

unusual quantity of moisture or inert material, or when an
instantaneous upset occurs. The significant thermal inertia
or "flywheel" effect of the mass of bed media, with the potential for dampening spikes and other transient phenomena, provides for inherent system control. Further, the thermal
inertia of the bed facilitates moderating the operating problems caused by instantaneous "slugs" of unusual feed materials.
These combustion control characteristics are highly significant.
    These advantages are in addition to the significant capi-
tal investment savings created by the simpler fluidized bed
incineration systems. The CFB can avoid an afterburner or
acid gas scrubbing system. It is recognized that fluidized
bed reactors cannot handle whole drummed waste or materials
with a high potential for slagging. Further, it is recognized
that fluidized bed reactors must handle solids that are sized
appropriately for the fluidization process. Consequently,
they are somewhat limited, particularly when compared to
rotary kilns. In future years, however, fluidized bed reactors will play an increasing role in centralized hazardous
waste incineration facilities, perhaps as companions to rotary
kilns.

IV. CONCLUSION

Permanent installations for the incineration of solid
and semi-solid (sludge) hazardous wastes are an increasingly
important management tool in the overall treatment and disposal
arena. A wide range of technologies are available including
slagging and ashing rotary kilns, rocking kilns, bubbling and
circulating fluidized bed boilers, and other approaches such
as hearth furnaces and retrofitted industrial installations.
Of these technologies, rotary kilns have become the design of
choice due to feedstock and combustion condition flexibility.
However, fluidized beds are emerging as a competitive technology where feedstock conditions permit. Rotary kilns have
the advantages of system simplicity and lower temperature
operating conditions. Combustion is readily managed in these
installations. The basic considerations of temperature, time,
and turbulence are manipulated by the use of excess air and
by the incorporation of heat rejection/heat recovery devices
into the process train. Turbulence or mixing can also be
manipulated by reactor selection (i.e., fluidized bed).

Technological maturity and consequent system availability
can be demonstrated for both rotary kiln and fluidized bed
technologies. System and disposal costs are a function of
specific waste characteristics, waste volumes, technologies,
and regulatory climates. Permanent installations for either
general hazardous waste destruction, or for specific industrial application, are therefore readily available.

REFERENCES

1. Anon. 1987. Improving Incinerator Performance with
   Oxygen Enrichment. *The Hazardous Waste Consultant* 5(2):
   1-20 - 1-22.

2. Anon. 1988. Integrated Hazardous Waste Management
   Facility Opens Near Swan Hills, Alberta. *The Hazardous
   Waste Consultant* 6(1):1-1 - 1-5.

3. California Air Resources Board (CARB). 1982. Technologies for the Treatment and Destruction of Organic Wastes
   as Alternatives to Land Disposal. Sacramento, California.

4. Cegielski, J. M. 1981. Hazardous Waste Disposal by
   Thermal Oxidation. John Zink Company, Tulsa, Oklahoma.

5. Chang, D. P. Y. et al. 1987. Evaluation of a Pilot-
    Scale Circulating Bed Combustor as a Potential Hazardous
    Waste Incinerator. JAPCA 37(3):266-277.

6. Chemcontrol. Undated. Dealing with Hazardous Waste
    Problems Statement of Qualifications. Chemcontrol A/S.
    Copenhagen, Denmark.

7. Clay, D. R. 1986. Letter to Mr. George Wessman con-
    taining the Approval to Dispose of Polychlorinated
    Biphenyls (PCBs) in the GA Technologies Circulating
    Fluidized Bed Boiler. Transmittal includes results of
    trial burn.

8. Frankel, I., Sanders, N., and Vogel, G. 1983. Survey of
    the Incinerator Manufacturing Industry. *Chemical Engin-
    eering Progress* 79(3):44-55.

9. Ho, M. D., and Ding, M. G. 1987. Field Testing and
    Computer Modelling of an Oxygen Combustion System at the
    EPA Mobile Incinerator. Proc.: American Flame Research
    Committee International Symposium on Incineration of
    Hazardous, Municipal, and Other Wastes. Palm Springs,
    California.

10. Keitz, E. et al. 1984. Profile of Existing Hazardous
    Waste Incineration Facilities and Manufacturers in the
    United States (Final Report). Mitre Corp., McLean,
    Virginia for USEPA. Contract EPA/600-2-84-052.

11. Lauber, J. D. 1982. Burning Chemical Wastes as Fuels
    in Cement Kilns. JAPCA 32(7):771-776.

12. Oppelt, E. T. 1986. Hazardous Waste Destruction:
    Thermal Techniques will be Increasingly Used as Legal
    Restrictions on Land Disposal Take Effect. *Environmental
    Science and Technology* 20(4):312-318.

13. Oppelt, E. T. 1987. Incineration of Hazardous Waste:
    A Critical Review. JAPCA 37(5):558-586.

14. Rinker, T. L. 1986. Successfully Siting Hazardous Waste
    Management Facilities. Presented at the Eigth Canadian
    Waste Management Conference.

15. Schofield, W. R., and Vingris, R. E. 1988. Off-Site
    Incineration vs. On-Site. *Waste Age*. May 44-54.

16. Short, H. et al. 1988. Europe Opts for Incinerating
    Hazardous Waste. *Chemical Engineering* 95(3):33-39.

17. Staley, L. J., Richards, M. K., Huffman, G. L., and
     Chang, D. P. Y. 1987. Incinerator Operating Parameters
     Which Correlate with Performance. Hazardous Waste
     Engineering Research Laboratory, USEPA, Cincinnati,
     Ohio.

18. Tang, J., and Taylor, E. S. 1986. The Ahlstrom Pyroflow
     Circulating Fluidized Bed Combustion System. *In* AFBC
     Technical Design Data Source Book (S. Tung, ed.) Pyro-
     power Corporation, San Diego, California.

19. Tillman, D. A., Seeker, W. R., Pershing, D. W., and
     DiAntonio, K. 1988. Converting Treatibility Tests into
     Conceptual Incineration Designs. Presented at the
     Combustion Institute Western States Meeting, Salt Lake
     City, Utah.

20. Various. 1984. Reviews of the Slagging Waste Incinera-
     tion Facility of Kommunekemi, Nyborg, Denmark. Reprints
     of articles published in the News Tribune, Woodbridge,
     New Jersey.

21. Vogel, G. A. et al. 1987. Incinerator and Cement Kiln
     Capacity for Hazardous Waste Treatment. Hazardous Waste
     Engineering Research Laboratory, USEPA, Cincinnati, Ohio.

Chapter VII

MOBILE, TRANSPORTABLE, AND DEVELOPING

INCINERATION SYSTEMS

I. INTRODUCTION

In response to the Superfund program (CERCLA), the Department of Defense program (DERA), and other federal and state regulatory efforts and the identification of thousands of sites contaminated with hazardous wastes, there has been an explosion in the field of site remediation. Many of the sites include soils and sludges contaminated with hazardous organics. Typical sites include military bases, industrial complexes, solvent recyclers, and many others. Incineration has been and is being considered to be one of few practical solutions for total site remediation, particularly when many contaminants are found at the same site. This can also be seen in EPA Record of Decisions (RODs). As a result, there has been a growing need for transportable and mobile hazardous waste incineration systems.

Transportable systems are relatively large facilities with capacities of 5-20 tons/hr that are shipped on up to 60 trailers. Transportable incinerators may require several months to be set up. They also require foundation work. The transportable systems discussed here include the ENSCO rotary kiln, the IT Corporation rotary kiln, and the OGDEN fluidized bed system. These are representative of systems offered not only by the suppliers mentioned but also by Chemical Waste Management, Roy F. Weston, John Zink Services Company, and several other firms. Mobile incinerators have smaller capacities and include systems that are contained on a few trailers that can quickly be mobilized. Mobile systems discussed here include two proven technologies: The Shirco infrared system and the Vesta rotary kiln. For each of the above mentioned systems, s system description and a disucssion of recent relevant experience at hazardous waste sites is presented. Developmental technologies are also discussed. There are many other systems in use and under development that are also worthy of mention. A list of typical system vendors and cleanup service companies is provided in Tables I and II.

TABLE I. Representative Manufacturers of Transportable
and Mobile Incineration Systems

Boliden-Allis (Allis-Chalmers)	Int'l Waste Energy Systems
C-E Raymond	John Zink Services
Detoxco	Kennedy Van Saun
Ensco/Pyrotech	M & S Engineering and Mfg. Co.
Environmental Elements	McGill Inc.
Ford, Bacon, and Davis	Ogden Environmental Services
Fuller Co.	Shirco Infrared Systems
J. M. Huber Corp.	Thermal
Industronics	Westinghouse

Source: C. R. Brunner [3].

TABLE II. Representative Suppliers of Transportable
and Mobile Incineration Site Services

Chemical Waste Management	O. H. Materials
Ecova Corp.	Rollins Environmental
Ensco	Riedel Environmental
Envirite Field Services	Tyger Construction Co.
Haztech Inc.	Vesta Technology, Ltd.
J. M. Huber Corp.	Waste Tech Services
International Technology	Westinghouse
Kimmins Environmental	Weston
Ogden Environmental	

It should be noted that these systems all are designed to manipulate the combustion mechanism elucidated in Chapter V. All are designed to achieve the regulatory requirements of 99.99% destruction and removal efficiency (DRE) for RCRA wastes, and 99.9999% DRE for TSCA wastes. Further, the transportable and mobile rotary kilns and fluidized bed incinerators follow the principles discussed in Chapter VI, including oxygen enhancement of the combustion air. The primary differences are in scale of operation as measured by capacity, and consequent foundation requirements. The following survey of systems is presented in that context.

II. TRANSPORTABLE INCINERATION SYSTEMS

Transportable systems have largely resulted from Superfund on CERCLA type work. They meet the need for systems to clean up hazardous waste sites with 10,000 to over 100,000 tons (9,100 to 45,500 tonne) of contaminated materials where incineration is the hazardous waste cleanup technology of choice. In such applications, the contaminated material is to be treated by incineration, delisted by passing either the EP-Toxicity or TCLP test, and then ultimately disposed of on the CERCLA site where it originated. Transportable systems typically have capacities of 5 to 20 tons (4.5 to 18 tonne) per hour of contaminated waste. Consequently, they require substantial support systems for waste feed, ash handling, waste heat rejection, air quality control, analytics (laboratory), and system controls.

Most transportable systems are ashing rotary kilns, conforming to the principles discussed in Chapter VI. They have the same L/D parameters, gaseous velocity constraints, cross sectional loadings, and related parameters. In addition to rotary kilns, transportable systems may employ low temperature volatilization technology and circulating fluidized bed technology. The dominance of rotary kiln technologies among transportable systems is significant, however, as illustrated by the ENSCO and IT systems.

A. The ENSCO Rotary Kiln System

The ENSCO transportable hazardous waste incineration system is one of the most proven technologies. It has been in service for several years. It is sufficiently large that it requires 20 trailers to transport to any site.

1. *The ENSCO Design*

The ENSCO design is based on the use of an ashing rotary
kiln and afterburner. Other systems include a waste heat
boiler and an air quality control system consisting of a
quench elbow, packed tower, ejector scrubber, and demister
[6, 10, 22, 7]. Refer to Chapter VIII for a discussion on
ejector scrubbers. A simplified schematic representation of
the ENSCO system is presented in Fig. 1. A picture of the
system is provided in Fig. 2. System components are discussed below.

a. *The ENSCO Rotary Kiln and Afterburner.* The rotary
kiln is 30 feet (9.1m) long with an inside diameter of 5.5
feet (1.7m), 7-foot 6-inch (2.3m) outside diameter, giving a
L/D ratio of 5.45. The unit is constructed of carbon steel
and lined with 6 inches (15.2cm) of refractory brick. The
kiln can be operated at any angle up to 6° and at a rotational
speed of 1/2 to 4 rpm. The kiln is of the cocurrent design
with the wastewater feed, clean and waste fuel feed, sludge
feed, and solids feed all at the burner end. The rotary kiln
discharges into a water quench, while the combustion gases
pass through a pair of refractory lined cyclones. The rotary
kiln accepts solid feed which is 4 inches minus. Solids are
fed with a variable speed 12-inch (30cm) diameter auger.
Typical kiln operating conditions are shown in Table III.

TABLE III. Operating Conditions of the Ensco Rotary Kiln

Parameter	Value
Thermal loading	15×10^6 Btu/hr (15.8 GJ/hr)
Outlet gas temperature	1400 - 1800°F (760 - 980°C)
Kiln stoichiometry	1.1 - 1.5
Solids residence time	30 - 60 minutes
Gas residence time	2 - 5 seconds
Incline	3 degrees
Rotational speed	1 - 3 rpm
Solids feed, approx.	1 - 6 tons/hr (0.9 - 5.4 tonne/h)

Source: ENSCO [6].

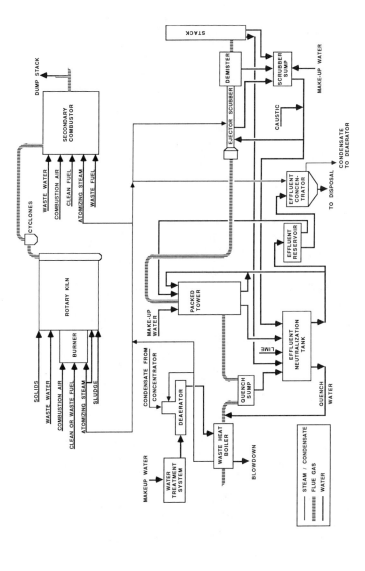

FIGURE 1. Simplified schematic representation of the ENSCO Model MWP-2000 transportable rotary kiln incinerator.

FIGURE 2. Pictorial view of the ENSCO MWP-2000 rotary kiln incinerator operating at the U.S. Air Force Base at Gulfport, Missouri. (Courtesy of Environmental Services Company, Little Rock, Arkansas).

The afterburner, or secondary combustion chamber (SCC), is mounted horizontally and is constructed of carbon steel with 4.5 (11.4 cm) inches of fire brick and 2.25 inches (5.7 cm) of insulating brick. The afterburner is 40 feet (12.2 m) long with an internal diameter (ID) of 6 feet 7.5 inches (2.0 m). The purpose of the afterburner is to incinerate the volatilized organics from the rotary kiln in an oxygen rich environment. Wastewater and waste fuel can be combusted in the afterburner. Typical afterburner operating conditions are presented in Table IV.

b. *The ENSCO Waste Heat Boiler and Quench System.* From the afterburner the combustion gases enter the waste heat boiler. The waste heat boiler includes a deaerator, water treatment system, and a quench prior to the air pollution control equipment. Steam produced by the boiler is used in the ejector scrubber for particulate removal. The boiler produces approximately 20,000 lbs/hr (9090 kg) of 250 psig (17 atm) steam for use in the ejector scrubber and for effluent concentration. The gases enter the waste heat boiler at >1800°F (980°C) and exit at 400–600°F (205–315°C). Gas velocities are maintained high to avoid particulate deposition on the boiler tubes. However, depositions could remain a problem when incinerating wastes with a high salt and volatile metal content. The gases exiting the boiler, enter a quench elbow where the temperature is reduced to saturation, near 165°F (74°C). The quench elbow is made of Inconel.

c. *The Air Quality Control and Auxiliary Systems.* The air quality control system accepts quenched stack gas and subjects it to multistage processing. The gases leaving the quench system enter the packed tower for removal of acid gases, followed by the ejector scrubber for removal of particulate and additional acid gases. The packed tower is 14 feet (4.3 m) high with a diameter of 6 feet (1.8 m). The

TABLE IV. Operating Conditions for the Ensco
Incineration System Afterburner

Thermal loading	20×10^6 Btu/hr (21.1 GJ/hr)
Outlet gas temperature	2200 – 2400°F (1200 – 1320°C)
Afterburner stoichiometry	1.2 – 1.5
Gas residence time	>2 seconds

Source: ENSCO [6].

unit is fabricated of fiberglass reinforced plastic (FRP) and includes an internal demister. Following the packed tower, the gases enter the ejector scrubber and finally a demister section and exits the stack. The ejector scrubber produces a sufficient pressure drop to act as the prime mover for the entire system, as a result an induced draft fan is not required. Excess steam that is produced can be vented to the atmosphere. The exhaust stack is also manufactured of FRP.

Other equipment in the ENSCO transportable incineration system includes the control room and material handling equipment. The type of material handling equipment depends on the nature of the contaminated medium onsite. The ENSCO system requires approximately 2 acres (8100 m^2) of space. The system is transported on 20 trailers to the site. The equipment is either skid or trailer mounted. The unit requires 30 to 40 days to be set up. The kiln is rebricked at each hazardous waste site.

2. The Experience of Using the ENSCO Rotary Kiln

ENSCO is one of the major services companies providing transportable hazardous waste incineration technologies to CERCLA sites. ENSCO has successfully remediated hazardous waste contamination at numerous locations. A brief summary of their experience is presented below.

In 1985 the ENSCO system was used to thermally treat 10,000 cubic yards (7645 m^3) of contaminated soils and sludges at Sydney Mines near Tampa, Florida. In 1986 and 1987 the unit was used for test burns at Gulfport, Mississippi. DREs of 99.99999% were achieved on polychlorinated biphenyl oil (PCB), dioxin surrogates, hexachlorethane, and trichlorobenzene. The removal of hydrogen chloride ranged from 99.4 to 99.9%, exceeding the requirements imposed by RCRA. Particulate removal ranged from 0.0083 to 0.030 grains/sdcf (0.019 to 0.06g/m^3) at 7% oxygen, well below the RCRA limit of 0.08 grains/sdcf (0.18g/m^3) at 7% oxygen. Principal Organic Hazardous Constituents (POHCs) were not detected in the stack gases down to the detection limits.

The ENSCO system is also being used for site remediation at the Lenz Oil Services site in Illinois, south of Chicago. This site includes contaminated soil, drums, liquids, and sludges. The major contaminates include volatile and semivolatile organics and chlorinated compounds. Clean-up goals for the soils are levels of 5 ppm of total organics for material with less than 50,000 ppm of total hydrocarbons. For material containing greater than 50,000 ppm of total

organics the remediation goal is the lesser of 2 ppm or a
10,000 fold reduction. Trial burn POHCs selected include
carbon tetrachloride and perchloroethylene.

B. The IT Corporation Transportable Kiln System

International Technologies (IT) Corporation is a major
operator of transportable incineration services. The following provides a system description and experience of the IT
Corporation transportable Hybrid Thermal Treatment System
(HTTS) using a rotary kiln [1, 11, 5].

1. The IT Transportable Incineration System Design

The HTTS is based on the use of a rotary kiln, afterburner, water quench, Hydro-Sonic wet scrubber (see Chapter
VIII), mist eliminator, and an induced draft fan. The system
is shown schematically in Fig. 3, and pictorially in Fig. 4.
The components of this incineration process are discussed
below.

a. *Rotary Kiln*. The HTTS rotary kiln is almost unique
in that it is of the countercurrent design rather than the
conventional cocurrent design. The waste feed and the heat
source or burner are at opposite ends of the kiln (see
Chapter VI, Figs. 7 and 9). Further, the burner can be moved
up to 10 feet (3 m) into the kiln as a means of adjusting the
rate of heat transfer and the temperature profile of the unit.
Burner placement provides the opportunity for significant combustion control. The unit is operated in a controlled air
fashion with a minimum or no excess air. As a result, the
unit is capable of a high throughput with reduced levels of
particulate carryover. The rotary kiln is designed to accept
wastes in the form of solids, sludges, and liquids. The
rotary kiln has an outside diameter of 7.5 feet (2.3 m) and is
45 feet (13.7 m) long. The inside diameter of the kiln is
6.5 feet (2 m), which gives an effective L/D ratio of 6.9.
Typical kiln operating conditions are presented in Table V.
The IT Corporation rotary kiln was manufactured by Allis-
Chalmers Corporation, a firm with substantial experience in
the application of rotary kiln technology to mineral ore
roasting, coal gasification, and cement manufacture. The
design features of the IT kiln, particularly the high L/D
ratio, reflect the Allis-Chalmer design approach. Conventional L/D ratios are in the range of 3.0 - 4.0.

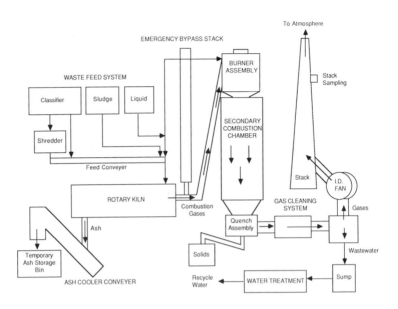

FIGURE 3. Schematic representation of the hybrid thermal treatment system of IT Corporation.

   b. Afterburner. The secondary combustion chamber for the IT system consists of a down-fired cylindrical shell lined with a castable refractory material. The burner is located at the upper end where the combustion gases are discharged from the kiln and auxiliary fuel are fed to the afterburner. Primary, secondary, and tertiary air are used to control combustion in the afterburner and to minimize the formation of NO_x. The unit can use fuel oil, natural gas, or light waste organic liquids as supplementary fuel. Contaminated aqueous liquids can also be fired in the afterburner for temperature control. Also, the afterburner can be operated as a stand-alone liquid incinerator.
   The lower section of the afterburner incorporates an integral quench chamber. Solids carryover from the kiln is removed from the quench chamber by a drag conveyor. Any suspended solids in the circulating water stream are removed by hydroclones. The effluent from the hydroclones is

FIGURE 4. Pictorial view of the IT Corporation HTTS (Hybrid Thermal Treatment System). Courtesy of IT Corporation, Knoxville, Tennessee.

TABLE V. Typical Operating Parameters of the IT
    Rotary Kiln

Parameter	Value
Thermal loading	4,000,000 – 30,000,000 Btu/hr (4.2 – 31.6 GJ/hr)
Outlet gas temperature	1000 – 1500°F (540–815°C)
Solids residence time	15 – 30 minutes
Kiln stoichiometry	Near starved air
Ash temperature	700 – 900°F (370–480°C)
Rotational speed	1 – 3 rpm
Soil feed rate	10 – 18 tons/hr (9–16 tonne/hr)
Burning firing rate	1 – 3.5 gpm oil (3.8 to 13.2 1/min)

Source: IT vendor data.

recycled back to the quench sump where it is used as spray
water to cool the combustion gases. Typical afterburner
operating conditions are in Table VI.

   c. *Gas Cleaning System*. The gas cleaning system includes
a gas conditioning chamber, a Hydro-Sonic Tandem Nozzle
Scrubber, a high efficiency mist eliminator, and a variable
speed induced draft fan. The fan discharges the flue gases
to a 60-foot (18 m) stack which has a continuous monitoring
system. The scrubbing system is designed to remove high
levels of acid gases and submicron particulate. High removal
efficiency of hydrogen chloride is important when incinerating
PCBs. The scrubber system uses sodium hydroxide (NaOH) as the
scrubbing media. Solids removal is incorporated into the
scrubber recirculation tank. The variable speed fan is used
during startup and for optimizing energy consumption during
normal operation. In addition to the use of the variable
speed controller flue gas can also be recirculated back to the
gas conditioner to increase the systems effective turndown
ratio. Recycling of the flue gas also helps to maintain
optimum gas velocities to the hydro-sonic scrubber.

   d. *Ancillary Systems*. Ancillary systems include the
waste preparation and feed system, wastewater treatment, and
a control room. In most cases the waste preparation and feed
system are enclosed in a temporary structure. This structure
is equipped with ventilation to control the release of toxic
gases and dust. Activated carbon is used to filter the air
before it is discharged from the building. Solids are fed
into a classifier to increase the homogeneity of the material.

TABLE VI.   Operating Parameters for the IT Afterburner

Parameter	Value
Operating temperature	2200°F (1204°C) for PCBs
	1800°F (982°C) for other hazard-ous materials
Gas residence time	2 seconds
Oxygen stoichiometry	3 percent oxygen
Thermal heat input	56,000,000 Btu/hr (51.9 GJ/hr)
Quench temperature	180°F (82°C)

   Source: IT vendor data.

A shredder is used to reduce the size of drums, fiber packs, rocks, tree stumps, and other material. The conveyors which feed the kiln are enclosed and ventilated.

A wastewater treatment system is used to treat any waters that are discharged from the incinerator system. Blowdown is collected from the quench and gas scrubbing systems. The pH of these waters is adjusted and suspended solids and heavy metals are removed. Soluble metals are reacted with NaOH to form insoluble metal hydroxide precipitates. These precipitates are removed by flocculation following polymer addition and finally by an incline, gravity plate clarifier. Following this the suspended solids concentration is further reduced to less than 5 ppm.

The system also includes a transportable control room providing centralized data acquisition and control. Analog and digital signals are fed from the various modules to the control room to provide input to both permissive and shutdown logic functions. Key process operating conditions such as stack emissions are stored on the computer.

e. *System Setup.* The incineration system is constructed of transportable modules. Each module is complete with all interconnecting electrical and piping and includes all pumps, pipes, tanks, instruments, and electrical equipment. Shipping dimensions were established to reduce the need for overweight and oversized loads. As an example the afterburner is constructed of four modules while the air pollution control equipment is constructed of five modules. A total of 56 trailers are required to ship the complete IT transportable incineration system and all ancillary equipment. At the Cornhusker Army Ammunition Plant (CAAP) site in Nebraska, it took four weeks for site work and required approximately 600 cubic yards (460 m^3) of concrete. At CAAP, the complete system was placed on existing foundations and ready for commissioning in twenty-five days. Approximately nineteen days after commissioning the unit was ready for operation.

2. *IT Incineration Experience*

The IT Corporation system was first demonstrated on nonhazardous material at the assembly site in Tulsa, Oklahoma. In 1987, a contract was awarded to incinerate 30,000 tons (27,200 tonne) of contaminated soil at CAAP. Here the soils are mostly contaminated with trinitrobenzene (TNB), trinitrotoluene (TNT), dinitrotoluene (DNT), and cyclotrimethylenetrinitramine (RDX). The soil was 60% sand, 25% silt, and 15% clay. The soil moisture content varied from 8 to 25% by weight. The DRE for TNT was greater than 99.9999% at a feed

rate of 15 to 17 tons/hr (14-15 tonne/hr). This exceeded
the RCRA requirement of 99.99% by two orders of magnitude.
The afterburner or SCC was maintained at 1600 - 1700°F
(870-925°C) during the tests. Particulate emissions ranged
from 0.0009 to 0.0028 grains/sdcf (0.002 to 0.006g/m^3).
Emissions of carbon monoxide ranged from 4 to 19 ppm. Overall
oxygen levels ranged from 6.6 to 7.2%. Hydrogen chloride
emissions ranged from 0.052 to 0.16 lbs/hr (24 to 73g/hr);
significantly below the RCRA limit of 4 lbs/hr (1810g/hr).
Following completion at the CAAP site the unit will be moved
to the Louisiana Army Ammunition Plant (LAAP) where the system will be used to treat 120,000 tons (109,000 tonne) of
similarly contaminated soils.

3. The Thermal Desorption System

IT Corporation is currently in the process of developing
a Thermal Desorption System (TDS) [12]. Presized soil is fed
continuously into a desorption furnace equipped with an
internal rotating cylinder. The rotating cylinder is indirectly heated and the flue gas is discharged to the atmosphere.
If natural gas is used, the exhaust gases would be relatively
clean. Since no air enters the rotating chamber, the quantity
of gases generated from the moisture and organics in the soil
is relatively small when compared to a conventional system.
The offgases can then be incinerated in an afterburner and
the resulting gases scrubbed before release to the atmosphere. A second option is that the volatilized gases can be
condensed and recovered. In the recovery mode, equipment is
provided to dispose of four distinct phases: vapors, organic
liquids, water, and sludge. A schematic representation of the
system is shown in Fig. 5. A pilot scale facility has demonstrated the TDS at a capacity of 2 tons/day (1.8 tonne/day).
IT is currently developing transportable designs for large
scale low temperature systems for use on soils.

C. The Ogden Environmental Circulating Bed Combustor

Ogden Environmental acquired the transportable incineration program of GA Technologies. This program employs a circulating fluidized bed incinerator which employs mixing and
turbulence, and maximized heat transfer, as the driving
mechanisms for thermal destruction (see Chapter VI). The
fundamentals of the CFB design are based upon extensive
laboratory, pilot, and full scale design experience of Hans
Ahlstrom Laboratory in Finland and Pyropower. Initial

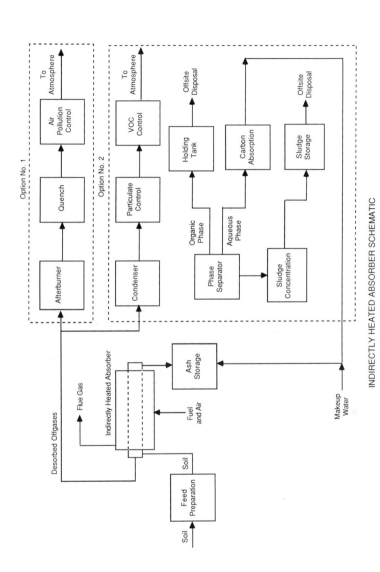

INDIRECTLY HEATED ABSORBER SCHEMATIC

FIGURE 5. Schematic representation of the thermal desorption system of IT Corporation.

results have demonstrated that this unique application of CFB technology has considerable promise for application to CERCLA and related type assignments. A pictorial view is presented in Fig. 6. System components and operating experience is presented below [26, 15, 16, 17, 18, 19].

FIGURE 6. Pictorial view of the Ogden circulating fluidized bed hazardous waste incinerator. (Courtesy of Ogden Environmental Services, San Diego, California).

1. System Description

The heart of the Ogden system is the ceramic lined CBC. The CBC is distinct from other fluidized bed combustors in that it operates at much higher velocities and as a result with much higher turbulence and carbon conversion. A schematic representation of the CBC is shown in Fig. 7. In the CBC solid hazardous waste and limestone enter the reactor and immediately are in intimate contact with the bed media of hot solids. Hazardous waste are typically injected directly into the sand and lime bed of the reactor. If soils are being burned, the soil itself becomes the bed media. Once the solids enter the bed, they are rapidly heated to reaction temperature, volatilizing the hazardous organics. A uniform reaction temperation (±50°F or 28°C) exists in the combustion chamber, the hot cyclone, and the recycle leg. Once the organics are liberated, they burn immediately in the reactor and as a result an afterburner is not required.

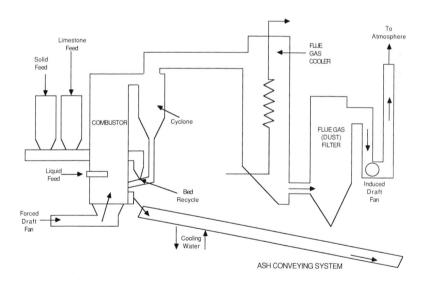

FIGURE 7. Simplified schematic representation of the Ogden circulating fluidized bed hazardous waste incinerator.

When solids are burned, they are recycled extensively through the system. Constant abrasion ensures the availability of fresh fuel for oxidation reactions. A particle may pass through the combustion chamber and hot cyclone 20 times before being discharged as fly ash or being removed with spent solids. Thus complete combustion is ensured in the form of contaminated soils (i.e., PCB soils), solids, or liquids.

Ash is removed by a water-cooled ash removal system. The ash contains calcium sulfate ($CaSO_4$ Gypsum) and calcium chloride ($CaCl_2$) from the in-bed removal of acid gases. As a result of in-bed acid gas removal, a wet scrubber is not required. The balance of the system includes gas cooling and particulate removal. Carbon monoxide and nitrogen oxide emissions are minimized by the availability of oxygen in the unit, the extensive turbulence and consequent mixing of reactants, the operating temperatures below the thermal NO_x temperatures, and secondary air injection in the combustion zone.

A heat injection system removes the thermal energy from the hot gases. Energy recovery occurs in both the combustion zone and the flue gas cooler. Sensible heat is further removed in heat exchangers placed between the cyclone and the fabric filter. Because the acid gases are removed, the materials of construction are subject to much less corrosion. Typical operating conditions for the CBC are presented in Table VII.

The largest of the transportable Ogden CFB unit uses a 36-inch (0.91 m) diameter combustion chamber. The complete unit measures 19 feet (5.8 m) by 33 feet (10 m) by 50 feet (15 m) high. The entire plant is broken down into transportable modules. System capacity as a function of bed size for

TABLE VII. Typical Operating Parameters of the Ogden CFB

Parameter	Value
Operating temperatures	<1600°F (871°C)
Steam pressure and temperature of heat rejection system	250 - 2600psig/250 - 450°F (17-177 atm/120 - 230°C)
Gas residence time	2 seconds
Solids residence time	30 minutes (<1 inch diameter)
Auxiliary fuel	None on wastes >2,900 Btu/lb (1.61 kcal/g) otherwise #2 oil or natural gas

Source: Ogden Environmental [19].

various waste streams is shown in Table VIII. As can be seen
the system capacity is maximized on soils containing organic
compounds.

2. Experience

The CBC reactor has been developed by numerous companies
including Hans Ahlstrom Laboratories, LURGI, and Gotaverhen.
In 1980, Hans Ahlstrom Laboratories and GA Technologies
formed Pyropower Corporation to supply circulating fluidized
bed systems to the United States boiler market. These systems
were designed to burn coal, peat, wood waste, municipal solid
waste (see Chapter IV), and oil. Many units are in operation
in the United States, as discussed in Chapters IV and VI, with
several more units planned for installation. Ogden is now
using the technology for both fixed and transportable hazardous waste incinerator services.

Currently Ogden has two transportable CFB hazardous waste
incineration units. The first unit will be used to clean up
approximately 45,000 cubic yards of PCB-contaminated soil at
the Swanson River oil production unit jointly owned by ARCO
and Chevron on the Kenai Peninsula of Alaska. The second unit
will be used in California. Unit numbers three and four will
be completed by January 1989 and March 1989, respectively.
The DREs measured for such transportable CFB units are shown
in Table IX.

The Ogden Environmental Transportable CFB, like the stationary CFB, has the advantages of few moving parts; combustion manipulation through turbulence, mixing, and highly
effective heat transfer; moderate combustion temperatures
(i.e., 1500-1800°F or 815-982°C) requiring less supplementary
fuel; in-bed control of acid gases; and system simplicity. The
transportable CFB requires no afterburner and no acid gas
scrubber. Consequently, it has capital cost advantages as
well. Currently only Ogden Environmental offers a transportable circulating fluidized bed. However, the potential advantages of fluidized bed incineration are sufficient that
more such units may be constructed and offered in the near
term. Further, transportable "bubbling" bed hazardous waste
incineration systems may also become more prominent.

D. Transportable System Conclusions

The emergence of transportable hazardous waste incineration systems has made incineration a more cost-effective
approach to the on-site remediation of hazardous waste sites.
High capacity systems such as the ENSCO and IT kilns have

TABLE VIII. CFB System Capacity as a Function of Reactor Inside Diameter

Waste type	Moisture (%)	Heat content Btu/lb (Kcal/g)	Throughput, tons/hr (tonne/hr)			
			24" ID (0.61m) reactor	36"ID (0.91m) reactor	60" ID (1.5m) reactor	104" ID (2.6m) reactor
Chlorinated sludge	80	1,330 (0.74)	0.5 (0.45)	1.1 (1)	3.1 (2.8)	9.5 (8.6)
Chlorinated liquids	4	7,610 (4.46)	0.2 (0.18)	0.5 (0.45)	1.5 (1.4)	4.4 (4.0)
Oil and solvent waste	13	11,230 (6.58)	0.1 (0.09)	0.3 (0.27)	0.9 (0.8)	2.7 (2.5)
PCB soils	10	0 (0)	1.2 (1.1)	3.0 (2.7)	–	–

TABLE IX. Destruction and Removal Efficiencies
Measured on the Ogden Transportable CFB Units

Waste type	DRE (%)
PCBs	99.9999
Pentachlorophenol	99.992
Malathion	99.9999
Dichlorobenzene sludge	99.999
Carbon tetrachloride	99.9992
Activated charcoal	99.97
Freon	99.9995
Trichloroethylene	99.9999
Chlorinated coke	99.99
Chlorinated organics	99.999
Solid aromatic nitrile	99.9999

improved the schedule and cost aspects of on-site incineration
substantially. Typical costs for transportable rotary kiln
hazardous waste incineration are shown in Table X [3]. Historical costs associated with transportable system incineration are shown in Table XI for cases when incineration is a
stand-alone function and when incineration is incorporated
into a complex site remediation program. As shown in Table
XI, incineration costs can range from $150 to $500 per ton
($165 to $550/tonne), and total remediation with incineration can cost up to $600/ton ($660/tonne) depending upon the
volume of material being incinerated, the type of contamination, and the concentration of contaminants.

III. MOBILE INCINERATION SYSTEMS

Mobile incineration systems complement the transportable
systems. Typically, mobile units require relatively few
trailers for transportation and can be installed easily on a
site. Mobile units may address a specialized segment of the

TABLE X. Representative Transportable Incineration
Costs (1988 $)

Activity	Cost	$/Ton
Mobilization/demobilization	$ 1,602,000	8
Site preparation	8,000	–
Excavation	2,713,000	14
Excavated material transport	533,000	3
Excavated material storage building	432,000	2
Fugitive dust control	3,129,000	16
General operations	990,000	5
Material handling	733,000	4
Lime basins	879,000	–
Incineration (labor, fuel reagent, etc.)	24,663,000	125
Water	60,000	–
Power equipment	480,000	2
Closure	81,000	–
Subtotal	$ 36,303,000	183
Fill	2,287,000	12
Residuals transport	6,390,000	32
Residuals disposal	30,004,000	152
Total construction cost	$ 74,984,000	379
Owner's cost		
Legal, administration, permitting (7%)	5,250,000	27
Engineering (5%)	3,750,000	19
Services during construction (8%)	6,000,00	30
Subtotal	$ 89,984,000	456
Contingency (20%)	18,000,000	91
TOTAL REMEDIATION COST	$107,984,000	548 (596/tonne)

Source: Brunner [3].

TABLE XI. Cost Estimates of On-Site Incineration
Assignments

Site	Material quantity tons (tonne)	Bid price $/ton (tonne) Total remediation	Incineration only
SCS, Del Ray, FL	5,000 (4535)	343[b](377)	-
Beardstown, IL[a]	7,500 (6805)	360 (396)	-
Sidney Mine, FL[a]	10,000 (9075)	550 (605)	170 (187)
Lenz Oil, IL[a]	12,000 (10890)	450 (495)	-
Florida Steel, FL	14,000 (12705)	-	250 (275)
Vertac, AR	20,000 (18150)	580[b](638)	-
Cornhusker, NE[a]	29,700 (26950)	450 (495)	135 (149)
LaSalle Electric, IL	30,000 (27220)	-	500 (550)
California	197,000 (178765)	530[c](583	120[c](132)

[a] Ash is expected to be disposed of on-site.

[b] Price excludes excavation costs.

[c] Engineer's estimate.

Source: Brunner [3].

incineration market, as is the case with infrared and gasification systems. Alternatively, mobile units may be small
capacity, generalized incinerators. Finally, some experimental mobile reactors have been developed. All such units
are represented in the discussions below.

A. Infrared Incineration (Shirco)

The Shirco infrared incineration system is based on the
use of an infrared heated primary chamber coupled to a conventional fuel fired afterburner. Shirco has recently filed
for bankruptcy under Chapter 11 provisions for reorganiza-
tion. The following discussion includes an infrared system
description and a discussion of its operating experience
[2, 4, 21].

1. *System Description*

The Shirco infrared system includes a primary chamber, afterburner, emergency bypass stack, venturi scrubber, ionizing wet scrubber, and a control van. A schematic representation of the system is shown in Fig. 8. Major system components are discussed below.

   a. *The Primary Infrared Chamber.* Contaminated soil enters the primary infrared reactor via a surge bin and screw feeder. The material is then spread and leveled before entering the primary chamber. The feed then falls onto a stainless steel belt conveyor where it is moved under infrared glow bars. The belt is supported by high temperature top and bottom rollers and shafts which penetrate the shell of the unit. The rollers and shafts are supported by externally mounted bearings. The primary chamber has a total of 54 glow bars consisting of two independent circuits of 27 bars each. Each circuit is capable of individual control. The glow bars are operated to maintain the proper thermal destruction temperature. For wastes capable of sustaining combustion, the glow bars are turned off;

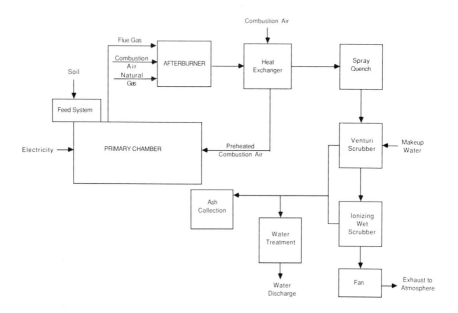

FIGURE 8. Schematic representation of the Shirco infrared systems incinerator.

however, they come on if the temperature decreases below a
given set point. The unit is operated at or below theoretical
air (equivalence of 1.0) requirements to minimize off gas generation. Combustion air is preheated in a heat exchanger to
lower the temperature of the exhaust gases from the afterburner. As the material moves along the stainless steel conveyor, it is agitated by rotating rakes, promoting good mixing. A blower at the discharge end of the unit establishes
countercurrent flue gas flow through the unit. Typical primary chamber operating conditions are presented in Table XII.

   b. Afterburner. The gases exiting the primary chamber
enter the afterburner where the volatilized organics are combusted. The afterburner is fired by four natural gas or propane burners mounted on the forward end of the vessel. Each
burner has its own independent air supply. It should be
remembered that the combustion gases exit the primary chamber
countercurrently. Typical operating conditions for the afterburner are presented in Table XIII.

   c. Gas Handling System. The gas handling system for the
infrared unit includes a spray quench section, venturi scrubber, and an ionizing wet scrubber. The gases exiting the
afterburner are cooled by water sprays to 300 to 400°F
(150-205°C) before entering a venturi scrubber. The

TABLE XII. Typical Operating Conditions for the
   Shirco Infrared System

Parameter	Value
Operating temperature	1650 to 1800°F (900 – 982°C)
Bed thickness	1 to 2 inches (2.5 – 5cm)
Pressure	–1 to –2 inches water column
Electrical energy	120 to 150 kW/ton of capacity
Feed material size	Best if <1.25 inches (3.2 cm)
Solid residence time	10 to 180 minues
Unit capacity	4 to 10 tons/hr (3.6 – 9.1 tonne/h) (combustible verses non-combustible materials like soil)
Unit dimensions	9 feet wide by 61 feet long (2.74 m wide × 18.6 m long)

Source: Vendor information.

TABLE XIII. Typical Operating Conditions for the
Shirco Afterburner

Parameter	Value
Operating temperatures	2200 to 2600°F (1204 - 1427°C)
Residence times	4 seconds for soils 2.5 to 3.0 seconds for combustibles
Unit dimensions	9 feet wide by 11 feet high by 72 feet long (2.7×3.4×22m)
Exhaust temperature	1800 to 2000°F (982 - 1093°C)
Fuel requirements	600,000 to 800,000 Btu/hr (633 - 844 MJ/hr)

Source: Vendor information

emergency bypass stack occurs before the gas quench section.
The venturi scrubber is constructed of fiberglass reinforced
plastic (FRP). The venturi scrubber operates at around 24
inches (61 cm) of water column and removes particulate down
to approximately 0.01 microns to a level estimated by the
author of 98%. Following the venturi the gases enter an
ionizing wet scrubber for acid gas and mist removal. The
ionizing wet scrubber and the venturi scrubber operate on the
caustic side. Following the ionizing wet scrubber the gases
pass through an induced draft fan and exit the stack. Wastewater flow is estimated from the unit at 30 gpm (114 1/min),
which can be reduced to 15 gpm (57 1/min) by ash quenching.
A typical water treatment system may include pH adjustment
and settling.

    d. *Design Analysis.* The mobile unit has a capacity of
80 to 250 tons/day (73-227 tonne/day) depending on the type
of material that is being incinerated. The primary chamber is
mounted on a single trailer while the afterburner and gas
handling system are mounted on two additional trailers. The
system also includes a support trailer and a control room and
laboratory trailer. No site preparation is required beyond
site grading. The trailers are leveled by hydraulic outriggers. Concrete pads can be placed under the outriggers
as required. Site requirements include 1 acre (4047 m^2) of
space, 1700KVA/480 volts, 15 amp/120 volt circuits, natural
gas or propane, a water supply, and a water discharge point.

The unit is operated by four personnel not including material handling.

The design, however, carries with it significant limitations. The contaminated solids feed in the primary reactor must be uniform, and thin, in order to ensure complete volatilization of the organics. The feed must be relatively dry, and should not contain free water. High moisture feeds make the unit volume limited due to the vaporization of moisture. Similarly high Btu wastes make the unit volume limited due to the space consumed by gaseous products of combustion. Volume limitations translate into reduced system capacities. These design limitations carry the implication that the Shirco design can be an effective, but specialized tool applicable only to select wastes.

2. *Experience With the Shirco Infrared Thermal Destruction Units*

The Shirco infrared technology has been used for industrial incineration, for thermal oxidizers, municipal sludge incinerators, carbon regeneration, and for many specialty applications. The Shirco infrared system has been tested at the Times Beach, Missouri, Dioxin Research Facility during the period July 8 to 12, 1985. The soil at Times Beach is contaminated with 2,3,7,8-tetrachloro-dibenzo-p-dioxin (TCDD). Test results as reported by Shirco for a feed of 227 ppb of TCDD and a 30-minute residence time indicated a DRE of 99.999996% and for a feed concentration of 156 ppb and a 15-minute residence time a DRE of 99.999989%. Particulate emissions were reported by Shirco around 0.001 grains/sdcf $(0.002g/m^3)$ at 7% oxygen. Tests on PCB contaminated soil obtained DREs of 99.9999% on feed ranging from 2560 to 2840 ppm. For these tests the operating conditions were as follows: 1470 to 1600°F (800–870°C) primary chamber temperature; average oxygen level of 8.6 to 12.2%; particulate emissions of 0.017 to 0.055 grains/sdcf (0.04 to 0.13 g/m^3) at 7% oxygen; and a waste feed of 60 lbs/hr (27.3 kg).

B. The Vesta Technology Mobile Incinerator

Vesta Technology, Ltd. was formed in 1986 to build and operate a mobile hazardous waste incinerator. Vesta's Thermal Destruction Units (TDU) are manufactured by Thermal Engineering, a subsidiary of Komline Sanderson Engineering. The mobile system is built around the use of a rotary kiln for the primary reactor. Vesta has two units: the Vesta 80

which is sized at 8,000,000 Btu/hr (8,400,000 KJ/hr) and has a
capacity of 0.5 to 1.0 ton/hr (0.45 to 0.90 tonne/hr) and the
newly developed Vesta 100 which has a capacity of 10,000,000
Btu/hr (10,550,000 KJ/hr) and can handle 1.5 to 2.5 ton/hr
(1.65 to 2.75 tonne/hr) of waste. The following is a discussion of the Vesta systems [23, 24, 25, 14, 9].

1. Systems Description

The Vesta systems include the rotary kiln as the primary
chamber coupled to a secondary combustion chamber (SCC). Combustion gases from the SCC enter a flue gas cooler, the dirty
gas side of a heat exchanger for flue gas reheat, a venturi
scrubber, a packed scrubber and demister, and finally the flue
gas reheat heat exchanger and the induced draft fan and
stack. A schematic representation of the Vesta 100 is shown
in Fig. 9 and a pictorial view of the unit in Fig. 10. The
design information on the Vesta 80 and 100 is confidential.
As a result only general information is presented here for
both the Vesta 80 and 100 systems.

   a. *The Vesta 80 Mobile Incinerator.* The Vesta 80 incinerator can handle liquids, solids, and sludge waste streams
containing hazardous organics. The system includes a rotary
kiln, secondary combustion chamber, gas cooler, a venturi
scrubber and separator, a packed tower scrubber, a demister
section, a stack gas reheat loop, induced draft fan, and
stack. The unit is mounted on a single trailer. The unit
can fire natural gas, liquefied petroleum, fuel oil, or suitable organic wastes. The unit requires 4 gpm of makeup water.
The rotary kiln is of the countercurrent design with an L/D
ratio of approximately 3. The burner is sized at 3,000,000
Btu/hr (3,200,000 KJ/hr). A horizontal screw conveyor is
used to discharge the ash. The SCC is of the cocurrent
design. Following the SCC the flue gases enter the flue gas
cooler, which is also used for flue gas reheat. From the gas
cooler the gases enter a three stage scrubber which includes
the venturi, separator, and packed bed scrubber. The scrubbing media is sodium hydroxide. Following the scrubber the
gas is reheated and exhausted to atmosphere through the
induced draft fan and stack. The system includes computer
controls, a feed system, and an ash handling system.

   b. *The Vesta 100 Mobile Incinerator.* The Vesta 100
includes a rotary kiln, SCC, three stage scrubber, an induced
draft fan, and a stack. The system is also mounted on a
single trailer. The kiln is of the countercurrent design
with an approximate L/D ratio of 5. The burner is sized at

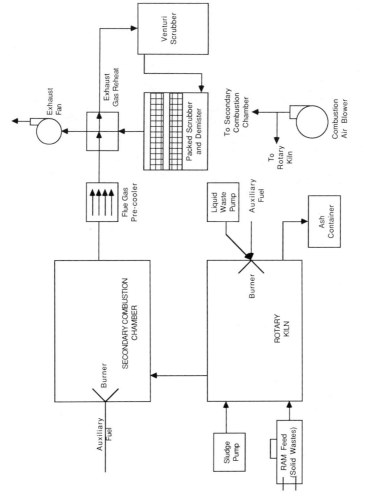

FIGURE 9. Simplified schematic representation of the Vesta hazardous waste incinerator.

FIGURE 10. Pictorial view of the Vesta Model 100 fully mobile hazardous waste incinerator. (Courtesy of Vesta Technology, Ltd., Fort Lauderdale, Florida.)

5,500,000 Btu/hr (5,800,000 KJ/hr). The kiln is coupled to an SCC. From the SCC the flue gas directly enters a low pressure drop venturi scrubber using a sodium hydroxide solution. Following this the flue gas enters a separator and a countercurrent packed bed scrubber and demister section. Again flue gas reheat is employed. Support systems are similar to above. The units can also be operated in a "multiunit" configuration with capacities of 6 to 8 tons/hr (5.5 to 7.3 tonne/hr).

2. Experience

Vesta has completed a test burn at "The Pit" in Aberdeen, North Carolina. The site is contaminated from illegally dumped and buried pesticides including Endrin, Chlordane, and Lindane. The Vesta 80 was used on the site fired with LPG. The DREs measured on incineration of the pesticides was 99.9998%. A DRE of 99.9999996% was measured for incineration of 10% concentrated carbon tetrachloride in soils. The unit incinerated 5,000 lbs (2270 kg) of material during the

test period. Particulate emissions were 0.018 grains/sdcf
(0.04 g/m^3).

   Another test burn was conducted at the Nyanza/Nyacol site
in Ashland, Massachusetts. The site is contaminated with
textile dye manufacturing residue. Vesta destroyed 1,200 to
1,600 tons (1,090 to 1,450 tonne) of sludge contaminated with
9,100 ppm of nitrobenzene. The DRE as measured was 99.9986%
incineration of nitrobenzene. Particulate emissions were
reported at a grain loading of 0.02 grains/sdcf (0.045 g/m^3).

C. Alternative, Developing, Mobile Incineration
   Technologies

   Other commercial and developing systems are discussed
below. The following literature was heavily used for this
section [13, 8, 20]. Again, space constraints limit the
technologies and the extent of the discussion.

1. Electric Pyrolysis

   The Electric Pyrolyzer is under development by Westinghouse Environmental Technology Division in Madison, Pennsylvania. Waste is introduced into the reactor by a rotary air
lock. The reactor has overhead and submerged electrodes in a
molten bed which supplies the energy for destruction of
organics in soil. Unit capacity is reduced if the waste
material has a substantial heating value. A 5 ton/day (0.45
tonne/day) mobile unit has been constructed and tested to a
DRE of 99199% on p-dichlorobenzene. The mobile unit is contained on two trailers. Westinghouse has plans to build a
100 ton/day unit (90 tonne/day).

2. Pyroplasma Process

   The Pyroplasma Process is being developed by Westinghouse
Plasma Systems in Ontario, Canada. The system includes a
plasma torch which is used for the destruction of liquid
hazardous wastes. The plasma is at a temperature of 18,000°F
(9980°C) with the reactor temperature maintained at 2500°F
(2245°C). Westinghouse has a prototype 1 gpm (3.8 l/min)
unit which has been tested to a DRE of 99.9999%. A mobile
1 gpm (3.8 l/min) and a 3 gpm (11.3 l/min) units are also
operational. Westinghouse is building four additional units.

3. Penberthy (Glassification) Pyro-Converter

The Penberthy Pyro-Converter is available from Penberthy
Electromelt International Incorporated in Seattle, Washington.
The furnace contains a molten glass pool at 2300°F (1260°C),
which is heated by submerged electrodes. Natural gas is used
for temperature control above the molten glass bed. The
technology has been tested up to a capacity of 150 tons/day
(136 tonne/day). The unit has achieved a DRE of 99.9999%
trichloroethylene. Currently, Penberthy Electromelt is
actively seeking markets in site remediation, asbestos
treatment, on-site incineration at industrial facilities, and
other markets.

4. Low Temperature Thermal Desorption

Roy F. Weston of West Chester, Pennsylvania, is developing an indirectly, hot oil heated, screw conveyor reactor to
be used for the thermal desorption of organics in soil. The
screw conveyor is heated to 450°F (230°C). The volatilized
organics are condensed and the oil fraction is recovered.
Wastewater is treated and generated volatile organics are
incinerated. The unit has achieved DREs of 99.99% on soils.
A transportable unit with a capacity of 7.5 tons/hr (6.8
tonne/hr) is currently under development.

5. Pyrolytic Incineration

Surface Combustion, Incorporated (formally the Surface
Combustion Division of Midland-Ross) in Toledo, Ohio, is
marketing the PyroTherm process. In this process hazardous
waste is continuously fed to a rotary hearth furnace, which
is indirectly fired. The furnace is heated to 1200 to 1600°F
(650 to 870°C) which is sufficient to vaporize the organics.
Solids residence times range from 5 to 30 minutes. The
chamber is operated in an oxygen free environment. The offgases go to a fume incinerator where they are destroyed at
1800 to 2000°F (980 to 1095°C) with a 1 to 2 second gas
residence time. The PyroBatch process is similar to the
PyroTherm process except that it operates in a batch mode.
Solids residence times range from 2 to 24 hours at a temperature of 1200 to 1600°F (650 to 870°C). A RCRA permitted
facility has been constructed at the McDonnel Douglas facility
in St. Charles, Missouri, to incinerate chlorinated hydrocarbons. The unit can accept four 55-gallon drums with an
8 hour processing cycle. Units are available in the size
range of 3,500 to 11,000 lbs/hr (1.6 to 5 tonne/hr).

6. Fast Rotary Reactor

PEDCo Incorporated of Cincinnati, Ohio, is developing a fast rotary reactor to incinerate hazardous waste. The rotary shell rotates at 10 to 30 rpm and includes internal flights for mixing of the solids. The reactor is divided into three distinct zones; the combustion zone, solid reheat zone, and the air preheat zone. In the solids reheat zone, the cool solids are reheated by the exiting combustion gases. The feed consists of hazardous waste, hot recycled solids (ash), and preheated air. Wastes with a heating value of as low as 1,650 Btu/lb (1740 KJ/lb) can be incinerated without the use of supplemental fuel. Limestone can be fed into the reactor to control acid gases. Low cost coal can be used for supplemental fuel as required. An 18 foot (5.5 m) by 2 foot 4 inch (0.7 m) rotary reactor has been constructed and tested.

7. Other Systems

Several other combustion/thermal destruction processes also are being developed for transportable and mobile use including molten salt incineration, plasma arc incineration of PCB capacitors, and a variety of other technologies. These are reviewed below.

Rockwell International of Canoga Park, California, is developing a molten salt incinerator for hazardous waste applications. The molten salt consists of sodium carbonate. Inorganics in the bed form oxygenated salts which are tapped and removed. Sizes are available up to 2000 lbs/hr (0.9 tonne/hr). The units have been tested on PCBs and achieved DREs of 99.9999%. Rockwell is currently seeking a firm to commercialize the technology.

Arc Technologies of Wayne, Pennsylvania, is developing a plasma arc process for the destruction of whole PCB capacitors. The prototype unit has a capacity of 1.5 tons/hr (1.36 tonne/hr) and operates at 3000°F (1650°C). The DREs on PCBs have exceeded 99.9999%. A prototype unit has been built at Chemical Waste Management's facility in Model City, New York. The unit will soon begin trial burns under the Toxic Substances Control Act (TSCA).

A listing of other developing or commercial technologies is presented in TABLE XIV. These technologies are only beginning to emerge in the incineration field.

There are many transportable and mobile hazardous waste incinerators in use today for site remediation. Major systems available include those using the rotary kiln as the

TABLE XIV. Developing Technologies for Hazardous
Waste Incineration

Developer	Technology
Applied Energetics, Inc. Tullahoma, Tennessee	Plasma incinerator
Battelle Columbus Laboratories Combustor, Columbus, Ohio	Multisolid fluidized bed
College Research Corporation Germantown, Wisconsin	Indirectly fired rotary kiln
Cottrell Environmental Science and Efthimion Enterprises Sumerville, New Jersey	Microwave plasma generator
FBD/BKMI Industrial Waste Systems, Salt Lake City, Utah	Pyrolyzing rotary reactor
Institute of Gas Technology Chicago, Illinois	Cycling cyclone incinerator
J. M. Huber Corp. Borger, Texas	Fluid wall reactor
Linde Division of Union Carbide, Tarrytown, New York	Linde oxygen combustion system
Rockwell International Denver, Colorado	Low temperature fluidized bed reactor
SCK/CEN Mol, Belgium	High temperature slagging incinerator
SKF Plasma Technologies Avon, Connecticut	Plasma dust process
Thermolytic Corp. Hercules, California	Thermolytic detroxifier
Waste Tech Services Golden, Colorado	Fluidized bed incinerator
Westinghouse Hanford Richland, Washington	Fluidized bed calciner

primary chamber. Such systems include those of IT Corporation, ENSCO, Roy F. Weston, and VESTA. Other systems being used include the fluidized bed unit of OGDEN. Fluidized beds have several advantages including lower operating temperatures, no need for an afterburner, and the use of in-bed removal of acid gases. The authors believe that there will be other fluidized bed systems commercially available in the near future. Another area where growth is expected is in the use of low temperature volatilization in the primary chamber of incinerators. Such systems are being developed by IT Corporation and Roy F. Weston. The primary chamber is used to volatilize the material where it is then destroyed in the secondary chamber. Options to destruction in the SCC include solvent or liquid recovery.

There are many other systems available and under development as can be seen in the above discussion. Each of these technologies will find a niche in the field of incineration. For example, melter technologies may find application for asbestos remediation, the destruction of low level waste, and incineration of materials that may require a stabilized ash product. Plasma technologies may find use in incineration of contaminated liquids, solvents, and PCBs. Pyrolytic incinerators may find continued use for on-site incineration at commercial and industrial facilities. Other technologies are unique to what is available and may find use in the future once their merit is recognized. Representative technologies include the PEDCo fast rotary reactor and the CORECo indirect fired rotary kiln. The field of transportable and mobile hazardous waste incineration is rapidly growing and changing. However, the dominant technologies of rotary kiln and fluidized bed have sufficient advantages (i.e., feedstock flexibility) to ensure broad competition in the near future.

REFERENCES

1. Accident, J. L., Burton, D. C., and Lovell, R. J. 1988 Cornhusker Army Ammunition Plant Remediation Project Case History. Proc: The International Conference on Incineration of Hazardous, Radioactive, and Mixed Wastes. San Francisco, California, May 3-6.

2. Berdine, S. 1987. Hazardous Waste Treatment Capabilities of the Shirco Infrared Demonstration and Full Scale Mobile Waste Processing System. Proc: The First Annual Hazardous Materials Management Conference of Canada. Metro Toronto Convention Center, Toronto, Ontario, September 9-11.

3. Brunner, C. R. et al. 1988. Incinerators for Site Clean-Ups: What's the Cost. *Waste Age,18*(5):55-60.

4. Brunner, C. R. et al. 1988. Site Remediation by Incineration/Thermal Treatment. Proc: The International Conference on Incineration of Hazardous, Radioactive, and Mixed Wastes. San Francisco, California, May 3-6.

5. DeCicco, S. G., and Accident, M. L. 1987. Transportable Hybrid Thermal Treatment System. Paper presented at the 24th AICHE/ASME National Heat Transfer Conference, Pittsburgh, Pennsylvania, August 9-12.

6. ENSCO. 1988. ENSCO Transportable Modular Incineration for On-Site Treatment of Hazardous Waste. Brochure, Remediation Services Division, Environmental Systems Company.

7. Frank, J. F. 1988. Illinois Environmental Protection Agency's Mobile Incineration Program--An Alternative Technology That's Working. Illinois Environmental Protection Agency.

8. Freemen, H. 1985. Innovative Thermal Hazardous Waste Treatment Processes. U. S. Environmental Protection Agency, Alternative Technologies Division, Cincinnati, Ohio, EPA/600/2-85/049, April 1985, 134 pgs.

9. Madeley, C. 1986. Aberdeen Pesticide Test Burn Appears Flawless to EPA. Moore County Citizen News-Record, Thursday, December 11.

10. McCoy and Associates. 1987. ENSCO's Mobile Incineration Configuration and Trial Burn Performance. *The Hazardous Waste Consultant,* 5(6):1-1 to 1-4.

11. McCoy and Associates. 1988a. IT's Hybrid Thermal Treatment System (HTTS). *The Hazardous Waste Consultant* 6(1):1-8 to 1-9.

12. McCoy and Associates. 1988b. IT's Thermal Desorption System. The *Hazardous Waste Consultant*, May/June, pgs. 4-2 to 4-3.

13. McCoy and Associates. 1988c. Special Features - A Guide to Innovative Thermal Hazardous Waste Treatment Processes. *The Hazardous Waste Consultant*, May/June, pgs. 4-1 to 4-43.

14. Nicholas, P. 1987. Nyanza Cleanup Ready to Begin. Middlesex News, Wednesday, September 30.

15. Ogden Environmental Services (Ogden). 1987a. In-Plant
     Waste Destruction with Ogden's Circulating Bed Combustion
     System . . . The Permanent Solution. Brochure, September.

16. Ogden Environmental Services (Ogden). 1987b. On-Site
     Clean Up Transportable Combustion System. Brochure.

17. Ogden Environmental Services (Ogden). 1987c. News
     Release . . . New Transportable Toxic Waste Disposal
     System Near Completion Says Ogden Environmental.
     October 30.

18. Ogden Environmental Services (Ogden). 1988a. News
     Release . . . Ogden Environmental Services, Inc.
     Awarded Cleanup Job by ARCO Alaska, Inc. April 15.

19. Ogden Environmental Services (Ogden). 1988b. Personal
     communication, June 6.

20. Radimsky, J., and Shah, A. 1985. Evaluation of
     Emerging Technologies for the Destruction of Hazardous
     Wastes. Prepared by Toxic Substances Control Division
     of the Department of Health Services, Sacramento,
     California. Prepared for U. S. Environmental Protection
     Agency, Alternative Technologies Division, Cincinnati,
     Ohio, EPA/600/2-85/069, June, 180 pgs.

21. Shirco. 1987. General Brochures from Shirco Infrared
     Systems, Dallas, Texas.

22. Sickels, T. W., and Lanier, J. 1988. Dioxin Incinera-
     tion in a Transportable Rotary Kiln Incinerator. Proc:
     The International Conference on Incineration of Hazardous,
     Radioactive, and Mixed Wastes. San Francisco, California,
     May 3-6.

23. Vesta. 1988a. Vesta 80 Technical Specification. Vesta
     Technology, Ltd. Fort Lauderdale, Florida, January 20.

24. Vesta. 1988b. Vesta 100 Technical Specification. Vesta
     Technology, Ltd. Fort Lauderdale, Florida, January 22.

25. Vesta. 1988c. General Brochures. Vesta Technology,
     Ltd. Fort Lauderdale, Florida.

26. Yip, H. H., and Diot, H. R. 1987. Circulating Bed
     Incinerators. Proc: The First Annual Hazardous
     Materials Management Conference of Canada, Metro Toronto
     Convention Center, Toronto, Ontario, September 9-11.

Chapter VIII

CONTROLLING PRODUCTS OF COMBUSTION

I. INTRODUCTION

This chapter discusses the control of products of combustion generated by the incineration of municipal solid waste (MSW) and hazardous waste. The types of emissions generated from such incineration or combustion systems are divided into three general categories: particulate matter, nitrogen oxides, and acid gases.

A. Particulate Matter

Particulate matter, also referred to as particulate, is the solid and/or condensable material that is carried over from the combustion process and either enters the boiler, heat recovery steam generator, or directly enters the air pollution control equipment. This material is commonly termed flyash once collected.

The quantity and characteristics of the particulate are a function of the combustion process and the type of material being incinerated. Parameters affecting particulate collectability in air pollution control equipment include particle size, thermal characteristics (i.e., ash fusion temperatures), hydroscopicity, resistivity, and specific gravity. Other condensable particulate is formed downstream of the combustor after the flue gases have had time to cool. Condensable particulate is composed mainly of hydrocarbons and usually condenses onto the fine particles in the gas stream. Typical particulate collection equipment includes baghouses (fabric filters), electrostatic precipitators (ESPs), and scrubbers. Cyclones and multiclone collectors may also be used in conjunction with the above equipment. However, their collection efficiency cannot achieve the requirements of the current regulations in most regions of the country. Consequently, this equipment is rarely used alone.

B. Nitrogen Oxides

Nitrogen oxides (NO_x) are formed by the oxidation of the
nitrogen in the combustion air and by oxidation of nitrogen
in the fuel or waste being incinerated. Emissions of NO_x can
be controlled using in-situ or post-combustion control technologies. In-situ control has been previously discussed and
includes staged combustion, low excess air combustion, reburn-
ing, and source separation of the waste. Post-combustion
control technologies discussed in this chapter include noncatalytic ammonia injection, catalytic ammonia injection (also
called selective catalytic reduction (SCR)), and flue gas
recirculation. Other systems also exist, including wet flue
gas denitrification. In the United States the control of NO_x
has become an increasing concern, as evidenced in the state
of California with regulations for proposed MSW and sludge
incineration facilities.

C. Acid Gases

Acid gases are formed from the sulfur, chlorine, and
fluorine present in the waste. Typical acid gases include
sulfur dioxide (SO_2), hydrogen chloride (HCl), and hydrogen
fluoride (HF). In hazardous waste incinerators, the generation of HCl could reach up to 5000 ppm from the incineration
of solvents containing very high levels of chlorine. At such
facilities, the halogenated wastes are carefully blended with
other wastes to reduce HCl loadings. Acid gas emissions at
hazardous waste incinerators have been controlled mostly by
wet scrubbing technologies. Recently, these types of scrubbers
have including condensing and noncondensing wet scrubbers such
as those designed by Gotaverken and Ceiba-Geigy. These scrubbers condense the flue gas to below 190°F (88°C) and in some
cases to as low as 90°F (32°C), well below the gas saturation
temperature. At this point, not only the acid gases are
removed but also many of the condensed PICs and some heavy
metals. Wet scrubber technologies are being considered for
several fixed hazardous waste incineration facilities in the
western United States.
Dry scrubbing is a technology being used extensively in
MSW facilities. Dry scrubbing involves spraying a slurry
into the gas stream where it reacts with the acid gases and
forms a solid residue. The typical system includes a spray
dryer and a baghouse, although in Europe and Sweden spray
dryer and ESP combinations have been used. Other acid gas
control systems include dry injection, a true dry process,

in which a solid reagent is injected into the gas stream
where it reacts with acid gases. Again, the material is
collected by a baghouse.

Emission requirements for certain states are presented
in Table I [6]. In the following sections, a discussion is
presented on the equipment used for the control of particulate,
nitrogen oxides, acid gases, and hydrocarbons. A discussion
is also presented on the characteristics of the collected
residue.

II. PARTICULATE CONTROL EQUIPMENT

Particulate matter (particulate) is generated when any
fuel or waste is combusted. Sources of particulate include:
entrainment of the ash present in the fuel or waste into the
flue gas, carryover of unburned carbon, products of incomplete
combustion (PICs), and the formation of soot particles and
condensable hydrocarbons. The type and quantity of particulate varies as a function of a host of variables including the
following (not exhaustive): type of combustor, type and
characteristics of the fuel or waste, level and distribution
of excess air between overfire and underfire, rate of combustion, and operation of the combustor.

This section describes the type of air pollution control
equipment available for the control of particulate. The discussion is limited to equipment currently being used on commercial and near commercial incineration facilities. Equipment is also discussed that is being used on mobile and transportable hazardous waste incinerators. As a result, this
section discusses baghouses (fabric filters), electrostatic
precipitators, and venturi scrubbers. Other, less conventional systems, are also discussed including: ejector scrubbers, porous ceramics, and high efficiency particulate air
(HEPA) filters.

A. Baghouses

Baghouses contain a densely woven fabric material through
which the gas stream passes. As the flue gas passes through
the material, the particulate is collected on the cylindrical
surface of the bag. To maintain the operation of the baghouse and to control pressure drop, it is necessary to remove
the collected particulate from the surface of the bag while
minimizing reentrainment. The method of particulate removal
varies with baghouse type. Important design factors include
the inlet and outlet grain loading (collection efficiency),
the gas flow and temperature, physical and chemical

TABLE I. Emissions Requirements for Resource Recovery Facilities in Selected States

Pollutant	New York Guidelines	California Guidelines	San Marcos Permit	Pennsylvania BAT Criteria	New Jersey Guidelines
Correction Factor	12% CO_2	7% CO_2	12% CO_2	7% O_2	7% O_2
Particulate gr/dscf (g/Nm^3)	0.02 (0.5)	0.01 (0.2)	0.01 (0.2)	0.015 (0.3)	0.015 (0.3)
Particulate <2 microns (g/Nm^3)	—	0.008 (0.02)	0.008 (0.02)	—	—
HCl	50 ppm/8 hr or 90% reduction	30 ppm	30 ppm/8 hr	30 ppm/1 hr or 90% reduction	50 ppm/1 hr or 90% reduction
HF	—	3 ppmv	3 ppmv/8 hr	—	—
SO_2	—	30 ppm	30 ppmv/8 hr	50 ppmv/1 hr or 70% reduction	50 ppmv/1 hr or 80% reduction
NO_x	—	140–200 ppm	200 ppmv/8 hr 425 ppmv/1 hr	—	300 ppm/1 hr 200 ppm/3 hr
Hydrocarbons	—	70 ppm	30 ppm/8 hr	—	case by case
CO	—	400 ppm	400 ppmv/8 hr	400 ppmv/8 hr 100 ppmv/4 da	400 ppm/1 hr 100 ppm/4 da

Source: Clark [6].

characterization of the particulate, required air-to-cloth ratio, and bag materials of construction. The most important baghouse design parameters are presented below.

1. Types of Baghouses

In a typical baghouse the flue gas is ducted to a manifold where gas is distributed to various compartments in the baghouse. Each compartment contains bags supported by a metal cage and mounted through a tube sheet. The flue gas is forced through the bags by the pressure differential across the baghouse, removing the particulate on the bag surfaces. The clean flue gas exits the bags and is sent to downstream equipment or exhausted to the atmosphere. Periodically, the bags are cleaned to remove the collected material. To accomplish cleaning, one compartment of the baghouse is isolated by the use of inlet, or both inlet and outlet, dampers.

Baghouses are designated by the method in which the bags are cleaned: shaking, pulse jet, and reverse air [27]. The shaker type baghouse collects particulate on the inside of the bag and requires a mechanically induced shaking action to release the collected material. The pulse jet baghouse collects the material on the outside surface of the bag and uses pulses of compressed air to release the collected material. The reverse air baghouse uses flue gas flowing in the reverse direction to collect the material. Although generally similar, the baghouses each have certain distinguishing features (Fig. 1). After the collected particulate is removed by cleaning, it is collected in a hopper where it is discharged by a rotary valve (or other type of air lock) to a conveying and handling system. Once complete, the compartment is again placed in service. This procedure continues for each compartment of the baghouse until the pressure drop is reduced to the design operating level. The whole sequence begins again when the pressure drop increases to a setpoint level.

The shaker type baghouse has the lowest capital cost but requires the most maintenance. Thus, the pulse jet and the reverse air baghouses are preferred for most applications. The pulse jet tends to be slightly less expensive than the reverse air, but requires compressed air for operation of the cleaning sequence. The reverse air generally has less bag wear than the pulse jet.

2. Design Factors

Key design factors necessary for consideration in the selection of a baghouse include the air-to-cloth ratio, particle size, fabric materials, and support equipment.

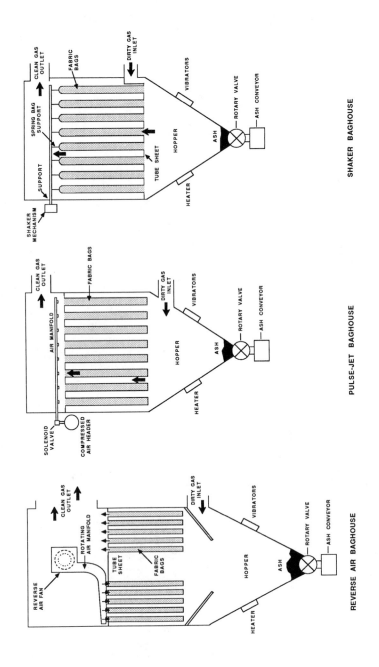

FIGURE 1. Schematic representation of a reverse air, pulse jet, and shaker type baghouse.

  a. *Air-to-Cloth Ratio.* The air-to-cloth (A/C) ratio is
the ratio of the flue gas flow through the unit in acfm
(actual cubic feet per minute) to the area of fabric available in square feet. The A/C ratio is calculated as the gross
or net ratio. The gross A/C ratio is calculated using the
total surface area of fabric in the baghouse, while the net
A/C ratio is calculated using the bag surface area that is
actually available for cleaning. For example, if a baghouse
has six compartments, one is usually off-line for cleaning;
the bag area of the compartment being cleaned is not used in
the calculation of the net A/C ratio. Typical net A/C ratios
used in combustion applications are 2:1 to 4:1. Note that as
the A/C ratio approaches 1:1, the size and cost of the baghouse increases. In ambient air baghouse operations (such as
for fugitive dust control) A/C ratios of as high as 10:1 are
possible.

  b. *Particle Size.* Particle size distribution and characteristics of the particle are very important for the design
of a baghouse. In order for a bag to properly collect particulate, it must build up a filter cake on its surface. This
filter cake is the actual surface which is collecting the
particles.   In this sense, the bag is acting as a support for
the filter cake. The larger the amount of small particulate
in the flue gas, the smaller the openings in the filter cake
and the finer the weave of the bag. As a result of this, the
pressure drop across the baghouse increases. Typical levels
of pressure drop are in the range of 2 to 5 inches of water
column (3.7 to 9.3 mm of mercury). The authors have observed
pressure drops as high as 10 to 12 inches of water column
(18.7 to 22 mm of mercury) on flue gas containing between 20
and 40% by weight of submicron particulate.

  c. *Fabric Materials.* Proper selection of bag fabric is
critical to the performance, operation, and maintainability of
the baghouse. Key factors for proper bag selection include:
ash composition, particle size distribution, flue gas temperature, moisture content, dew point of the flue gas, and the
hydroscopic behavior of the collected material. For combustion applications, bags are usually of a synthetic fabric
weave or a glass weave. These materials can be covered on
the dirty side by a tight drawn material such as Gortex. A
good example of a bag is Gortex supported on a Nomex backing
as shown in the electromicrograph in Fig. 2. Here the Nomex
is supporting the very fine Gortex material.
    Bag coatings such as Teflon improve the release of the
particulate from the bag's surface while fine membranes help

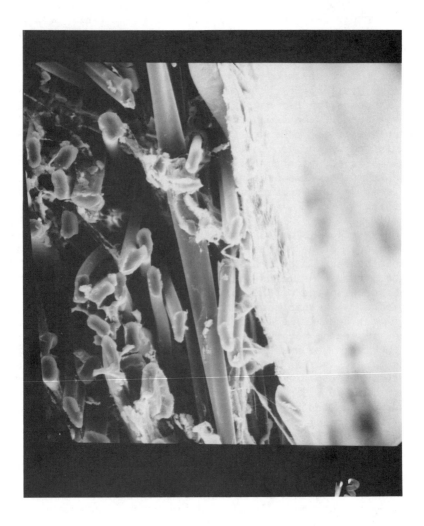

FIGURE 2. *Scanning electron micrograph of a Gortex membrane over a Nomex backing.*

improve the collection of fine particulate. Important bag design factors include permeability, mechanical strength, and heat resistance.

Permeability. The bag permeability is a measure of the bag porosity. It is defined as the air volume in cfm that will pass through a square foot of new clean cloth (usually ft^3/ min/ft^2 at 0.5 inch of H_2O or m^3/min/m^2 at 1.0 mm of mercury). Higher permeability makes it easier for flue gas to pass. It also results in a lower pressure drop and in most cases lower efficiency in collecting fine particles.

Mechanical Strength. The mechanical strength relates to the ability of a bag to withstand the strain of cleaning. For example, fiberglass bags should not be used on pulse jet bag-houses because they are easily damaged by the 100 to 120 psig (6.8 to 8.2 atm) header pressure during cleaning. This can be minimized to a certain extent by lower pulsing pressure or modified and more expensive bag cage designs.

Heat Resistance. The bag material must be able to with-stand the expected operating temperatures as well as any temperature excursions which could occur during upset condi-tions. Fiberglass bags work well up to the range of 400 to 500°F (200-260°C). Several metal woven bags for higher temp-erature operation are currently under development.

d. *Support Equipment*. Important baghouse support equip-ment includes the necessary ducting, isolation dampers, hoppers, insulation, mechanical supports, and controls. Also of impor-tance are the use of hopper heaters in cold climates and on ash that is very hydroscopic. Hopper vibrators are desired to avoid bridging. However, a hopper vibrator can turn an already bridged hopper into a compacted bridged hopper. The baghouse hoppers should be completely empty before the isolated compartment comes back on line. This serves two functions: (1) it reduces reentrainment of the collected particulate, and (2) it reduces hopper plugging. Baghouses and hoppers should be properly insulated to avoid condensation problems at start-up and shutdown. Preheating may be required for baghouses containing membrane coated bags. Such preheating can be done using preheated air or an in-line duct burner, but a bypass stack should be positioned before the baghouse as a safety precaution. For a hazardous waste incinerator, this may not be allowed or may require a gas flare to avoid the release of toxic PICs. A baghouse also needs a fire detection and con-trol system. A bypass duct is also used in many applications.

3. *Collection Efficiency*

Whether baghouses or ESPs achieve the greatest efficiency on the collection of particulate is still under debate. How-ever, the baghouse has several desirable features;

1. It presents a physical barrier to the particulate in the flue gas stream.

2. On spray dryer applications, it has been shown that
    the cake on the bag surface provides additional reac-
    tion for the removal of acid gases.
3. It better handles temperature excursions and changes
    in flue gas volume, which are typical of incinerator
    operation.

The collection efficiency of a baghouse can be measured
as the percent of the particulate removed or as the penetration, which is 100 minus the percent particulate removal.
Typically, baghouses can achieve collection efficiencies well
above 99.5% on particulate less than 10 microns in size
(respirable particulate) as shown in Fig. 3. It efficiently
collects submicron particulate which has been shown to have
high concentrations of heavy metals, condensed hydrocarbons,
as well as dioxins and furans when compared to other fractions. Particulate emissions at selected MSW facilities are
summarized in Table II. Table II demonstrates that the outlet grain loading for baghouses can reach a low of 0.006
grains/sdcf (0.014 g/Nm^3).

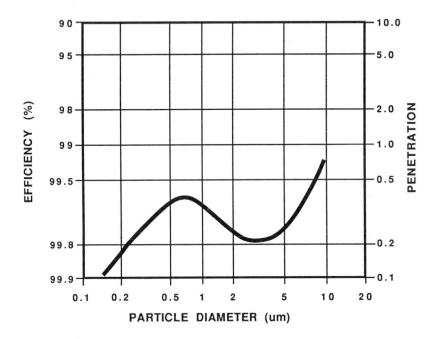

Figure 3. Typical fractional efficiencies for a fabric
filter. Adapted from [7].

TABLE II.   Particulate Emissions at Selected Facilities

Plant	Equipment	Particulate	
		gr/dscf	(g/Nm^3)
Bridgewater, MA	Fabric filter	0.020	(0.046)
Nashville, TN	Fabric filter	0.008	(0.018)
Framingham, MA	Dry scrubber/ fabric filter	0.020	(0.046)
Tsushima, Japan	Dry scrubber/ fabric filter	0.010	(0.023)
Wurzberg, Germany	Dry injection/ fabric filter	0.0012	(0.003)
Malmo, Sweden	Dry injection/ fabric filter	0.004	(0.009)
Kempton, Germany	Dry injection/ fabric filter	0.0016	(0.004)
Hogdalen, Sweden	Dry scrubber/ fabric filter	0.006	(0.011)
Hogdalen, Sweden	Dry injection/ fabric filter	0.006	(0.014)

Although ESPs dominated on MSW incineration facilities in
the 1970s, the trend has changed in many places. For example,
California, Washington, Connecticut, Michigan, and the Environment Canada prefer using baghouses to control particulate
in conjunction with spray dryers for acid gas control. The
above collection efficiencies should be achievable on fixed
hazardous waste incinerators. Baghouses are not used on
mobile and transportable systems, however, because of size
constraints associated with transportation.

B.   Electrostatic Precipitators

Electrostatic precipitators (ESPs) collect particulate
in the flue gas stream by electrically charging the particles, and under the influence of the electrical field separate
the particulate from the gas stream. An ESP consists of a
discharge system which charges the particles and a collecting
surface which collects the charge particulate. A simplified

sketch of an ESP is provided in Fig. 4 [31]. Particulate
that enters the ESP is charged by either a wire–weight system
or a rigid frame discharge electrode system. The wire–weight
system consists of suspended wires which are held taut by
suitable weights at the bottom end. The less commonly used
frame consists of a charged grid. Opposite and parallel to
the discharge electrodes are the collection electrodes. The
charged particules migrate to the collection electrodes where
they adhere and lose their charge. Like the baghouse, it is
necessary to remove the particulates from the collection
electrodes. This is done by mechanical vibrators or rappers
that are mounted on the collection electrodes. The collected

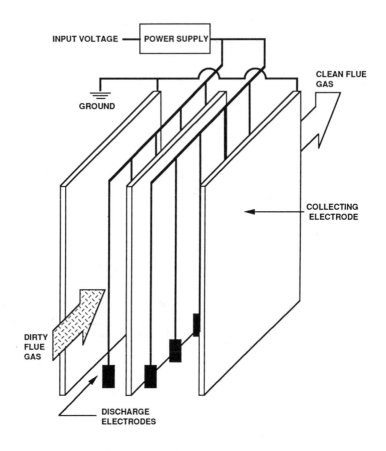

FIGURE 4. Simplified schematic representation of an
electrostatic precipitator. Adapted from Schneider et al [31].

particulate then falls into a collection hopper where it is
removed in a similar fashion to that discussed for the baghouse. Again, it is important to minimize reentrainment of
particulate during rapping.

1. Types of Electrostatic Precipitators

There are two types of ESPs: dry and wet . In the dry
ESP the collected particulate is rapped or vibrated from the
plate surface. In the wet ESP the collected particulate is
washed from the plate surface. Each type of ESP is briefly
discussed below.

a. *Dry Electrostatic Precipitators*. The dry ESP has
been the most common system used on MSW facilities. It
removes the collected particulate from the plate surface by
mechanical vibration. The size of an ESP is defined by its
number of fields. A field is the number of bus sections that
the unit has arranged in the direction of the gas flow. Most
older ESPs were only two-field units which proved to be less
effective. In general, a manufacturer will use one field for
up to 90% collection efficiency, two fields for up to 97%,
three fields for up to 99%, and four or more fields for collection efficiencies in excess of 99% [31]. The number of
bus sections arranged parallel to the flue gas flow is defined
as the number of cells. The tubular ESP which has been historically used in removing acid mists from ore roasting is
finding application on incinerators [30]. These units include
single and two-stage circular or rectangular tube designs. The
square tube, formed into a honeycomb structure, is the preferred design.

b. *Wet Electrostic Precipitators*. The wet ESP is similar
to the dry one except water sprays are provided to continually
wash the collection and discharge electrodes. A wet ESP is
necessary for material which cannot be rapped from the surface
of the collection plates as well as for the following reasons:
(1) collects acid mists; (2) used on high or low resistivity
ash; (3) provides a continuous cleaning action through the use
of water; (4) reduces reentrainment; (5) provides gas quenching and absorption; and (6) is compact in size [30, 31].
A two-stage, wet tubular ESP has been developed for incinerators [32]. The unit is of an upflow design and includes
the following operations in the direction of the gas path:
gas quenching to saturation with water, coarse screening of
particulate, acid gas acrubber, and the tubular ESP section.
These units can remove particulate to the RCRA limit of 0.08

grains/sdcf (0.18 g/Nm^3) to as low as 0.01 grains/sdcf (0.02 g/Nm^3) [32]. However, disadvantages include the necessity to collect and treat wastewater.

2. Design Factors

Proper design of an ESP includes the following factors: gas volume, operating temperature, dust characteristics, particle size, resistivity, and the required efficiency. Design parameters are briefly discussed below.

a. *Specific Collection Area*. The specific collection area (SCA) of an ESP is a measure of the amount of collection electrode area for every cubic foot (meter) of flue gas. The SCA is defined as the plate area in square feet (m^2) divided by the acfm (m^3). Typical values of SCA are around 0.3 (1.0 metric) for particulate with low resistivity to around 0.8 (2.6 metric) for particulate with a high resistivity. The higher the SCA the larger the physical size of the ESP. This results in reduced gas velocities and longer residence time in the collection area. Particulate concentration in the flue gas is a function of SCA (Fig. 5), and as the SCA increases, the particulate removal increases. Economic limits regarding an affordably sized SCA ultimately are reached.

b. *Resistivity*. The collection efficiency of an ESP is strongly influenced by the resistivity of the particulate. Particulates with low resistivity (less than 10,000 Ohm-cm) readily give up their charge to the collecting electrode and remain in the gas stream. Particulates with very high resistivity (greater than 10,000,000,000 Ohm-cm) do not relinquish any of their charge to the collection electrode and act as an insulator. Particulates with resistivity between these limits give up a portion of their charge and are collected on the collection electrodes. Resistivity changes significantly as a function of temperature in the range of 300 to 600°F (150 to 315°C). Typically, ESPs operate at temperatures around 450 to 550°F (230 to 285°C). At temperatures above 590°F (310°C), material can begin to fuse to the plates. The resistivity of MSW particulate changes rapidly in the range of 200 to 400°F (95 to 205°C), reaching a peak at around 300°F (150°C) and then rapidly decreases [4]. This change in resistivity may be even more significant for hazardous waste incinerators.

c. *Fields and Velocity*. Other important factors include the number of fields and the gas velocity. Particulate concentration in the flue gas is a function of gas velocity (Fig. 6). As the velocity of the flue gas through the unit

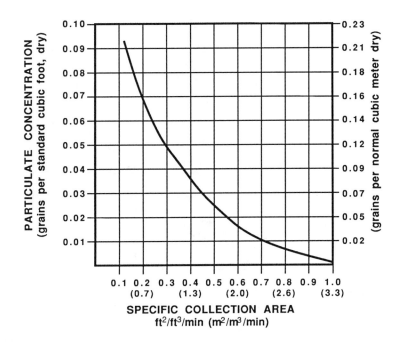

FIGURE 5. Specific collection area verses particulate concentrations for an electrostatic precipitator. Adapted from Clark [6].

decreases, the collection efficiency increases. The number of fields also plays a significant role in the collection efficiency.

3. Collection Efficiency

In theory, each field of an ESP collects 90% of the particulate. Three fields would have a collection efficiency of 99.9%. However, this is not the case since the remaining particulate passing a given field are smaller in size and as such more difficult to collect. The penetration and fractional collection efficiency for an ESP is not as good on smaller particulate as the baghouse (Fig. 7).

Operating conditions and grain loadings for ESPs at MSW incineration facilities range from 0.03 to 0.09 grains/sdcf (0.07 to 0.21 g/Nm^3) at 12% CO_2 (Table III). Newer installations at German plants have achieved values as low as 0.01

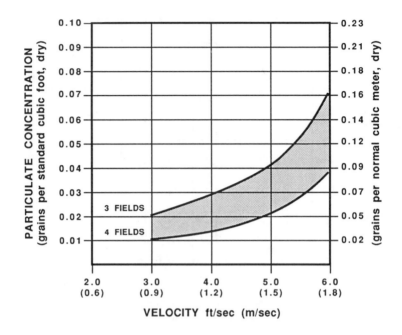

*Figure 6. Velocity and number of fields versus particu-
late concentrations for an electrostatic precipitator.
Adapted from Clark [6].*

grains/sdcf (0.02 g/Nm^3). The four-field ESP at Baltimore,
Maryland, is reported to have achieved 0.003 grains/sdcf
(0.006 g/Nm^3).

C. Venturi Scrubbers

Venturi scrubbers have found extensive use on hazardous
waste incinerators. They are not commonly used on MSW facilities because of sizeable pressure drops. A brief description
of venturi scrubbers is provided below.

1. *System Description*

Flue gas is injected into the narrow throat of a venturi.
The throat area is flooded to eliminate a wet-dry interface
where corrosion may occur. As the gas moves into the throat

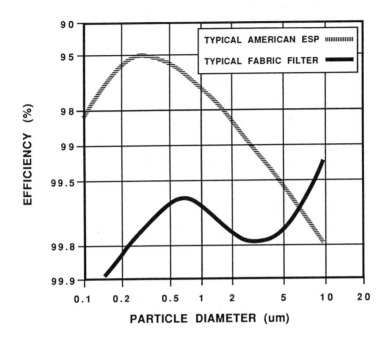

FIGURE 7. Typical fractional efficiencies for an
electrostatic precipitator. Adapted from EPRI [7].

area, it rapidly accelerates. The droplets of water form a
fine mist which entraps the dust particles by inertial impaction. As the gas leaves the throat area, it carries liquid
droplets that have reached velocities nearly equal to that of
the gas stream. The mixture then enters an expander section
where the gas is decelerated. The kinetic energy from the
liquid is transferred back to the gas stream due to drag
forces, resulting in the partial recovery of energy previously
lost when the gas accelerated in the throat area. Once the
mixture of gas and water has decelerated, it enters a cyclonic
separator for removal of the agglomerated particulates.

2. Collection Efficiency

The typical collection efficiency of venturi scrubbers is
a function of particle size and pressure drop (Fig. 8). As

TABLE III. Operating Parameters for Full Scale ESPs on MSW Waste-to-Energy Facilities

Location (Type)	Capacity Tons (tonne)/day		Input kVA	Temperature		Outlet gr/sdcf at 12% CO_2	
						gr/sdcf	(Nm^3)
Montreal, Canada Waterwall	1,200	(1,090)	35	536	(280)	0.08	(0.18)
Southshore, NY Refractory lined	500	(455)	33	600	(315)	0.06	(0.14)
Dade County, FL Refractory lined	300	(270)	48	570	(300)	0.03	(0.07)
Washington, DC Refractory lined	1,500	(1,365)	77	550	(290)	0.05	(0.11)
Quebec, Canada Waterwall	1,000	(910)	–	500	(260)	0.09	(0.21)
Saugus, MA Waterwall	1,200	(1,090)	–	390	(200)	0.03	(0.07)
Dade County, FL Waterwall	3,000	(2,730)	–	460	(235)	0.05	(0.11)
Pinellas, FL Waterwall	2,000	(1,820)	–	460	(235)	0.03	(0.07)

Source: California Air Resources Board [4].

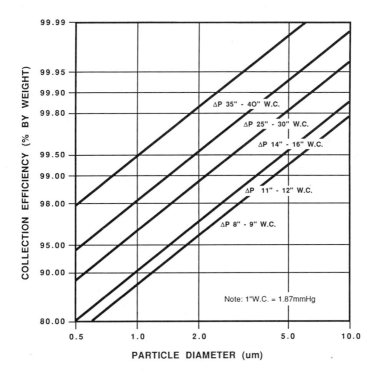

FIGURE 8.   Fractional efficiency curves for a venturi
scrubber. Source: Western Precipitation [36].

pressure drops increase, particulate removal increases. The
pressure drop selected for operation depends on the regulatory
requirements.

D. HEPA Filters

HEPA filters are used widely on incinerators which handle
radioactive wastes. They have a collection efficiency of at
least 99.97% on 0.3 micron particles [23]. Manufacturers have
developed filters with collection efficiencies up to an order
of magnitude below the required minimum. These filters con-
sist of corrugated separators in a wood or particleboard box.
The filters generally operate at ambient conditions and are

expensive to install and operate. However, they may be
required on facilities that thermally treat radioactive waste
such as melters, following scrubbing and cooling of the offgas
stream.

E. Porous Ceramic Filters

    A porous ceramic filter is a ceramic cylinder flanged at
one end and closed at the other. The units available are
80 (203 cm) inches long with an outside diameter of 2.375
inches (6.032 cm) and a wall thickness of 0.4 inches (1.016 cm)
[37]. These filter candles, as they are called, are mounted
on a tube sheet just like the bags of a baghouse; except the
enclosure is a pressure vessel. A schematic of a ceramic
candle filter is shown in Fig. 9. These vessels are

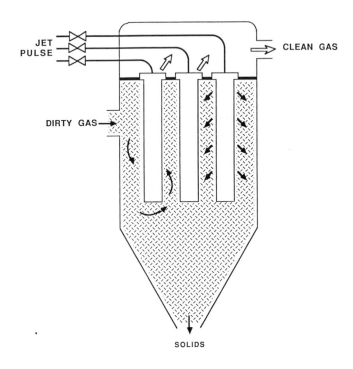

FIGURE 9. Schematic representation of a ceramic candle
filter: Source: Zievers, Eggerstedt, and Kulosek [37].

refractory lined and operate at pressures up to 170 psig
(12 atm). The porous ceramics can operate at temperatures
up to 2000°F (1090°C). Although expensive, this technology
can be applied to radioactive waste incinerators, small scale
plasma, and melter incinerator technologies.

F. Ejector Scrubbers

Ejector type scrubbers have been applied to many hazardous
and radioactive waste incineration processes. One type
described in this section is the HYDRO-SONIC scrubber currently
being used on several mobile hazardous waste incinerators.

1. *System Description*

The basic configuration of a HYDRO-SONIC scrubber (Fig. 10)
involves a steam ejector with water injected around and onto
the steam. The water is shattered into high speed droplets
which impact and encapsulate the dust particles [21]. The
mixture of steam, water, dust, and flue gas is driven down a
mixing tube where agglomeration occurs. The agglomerated
droplets are then separated in a cyclonic separator section.
Other means of powering the HYDRO-SONIC scrubber are
available including air driven systems, fan driven systems,
and a combination of fan and steam drive. In the steam-driven
system an induced draft fan is not required, since the ejector
supplies the operating pressure. Almost all large mobile or
transportable incinerators use an afterburner for destruction
of volatilized organics. The temperature of the offgas from
the afterburner is in the range of 2000 to 2300°F (1095 to
1260°C). As a result, a tremendous amount of heat is available
in the cooling gas. This heat can be used to generate steam,
and this steam can then power the HYDRO-SONIC scrubber. As
an alternative, the hot gases can be air quenched and a fan-
driven system used.

2. *Efficiency*

Collection efficiency and penetration is a function of
particle size for the HYDRO-SONIC scrubber (Fig. 11). The
unit is very effective on particulate less than 10 microns
and on submicron particulate, which has been shown to concentrate metals and dioxins. The system also has been shown to
remove HCl to a level of 99% and approximately 70% of the
inlet sulfur dioxide on MSW applications using a lime buf-
fered scrubbing water. Typical operating pressure drops for
fan-driven systems are in the range of 30 to 40 inches of
water column (56 to 75 mm mercury) (Shreveport Municipal

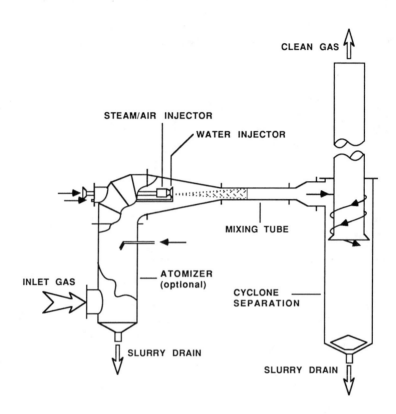

Figure 10. Drive configuration of a steam/air ejector scrubber. Source: Means and Holland [21].

Incinerator in Louisiana) which only required 0.10 grains/sdcf (0.23 g/Nm^3) corrected to 12% carbon dioxide.

Typical steam consumption ranges from 0.08 pounds (0.04 kg) of steam per pound of flue gas (lbs steam/lbs gas) to 0.6 lbs (0.27 kg) steam/lb of gas. Listed in Table IV are the typical grain loadings corrected to 12% CO_2 as a function of steam consumption. The above relationships are linear and other values can be calculated. For radioactive incinerators a two-stage steam drive can be used. Such a drive increases the collection efficiency approximately one order of magnitude (as indicated for the 0.30 and 0.60 steam usage values above).

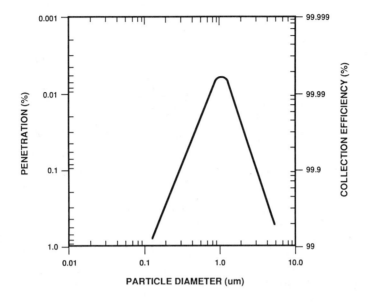

FIGURE 11. *Fractional efficiency calculated using impactor data for an ejector scrubber.* Source: *Means and Holland* [21].

TABLE IV. Emissions Rate as a Function of Steam Usage

Steam usage lbs steam/lbs gas	Emissions	
	Grains/sdcf	(g/Nm^3)
0.08	0.10	(0.23)
0.12	0.03	(0.07)
0.16	0.009	(0.02)
0.30	0.0002	(0.0005)
0.60	0.00002	(0.0005)

G.   Summary

In summary the major types of particulate control equipment include the baghouse and the ESP for MSW and fixed hazardous waste incinerator facilities. Mobile hazardous waste
incinerators use venturi scrubbers, wet ESPs of a horizontal
configuration, and ejector type scrubbers. As previously
indicated, cyclones and multiclone collectors were not discussed since neither can achieve the desired level of particulate control required by most regulations. A summary of
the collection efficiency of selected particulate control
devices (where data was available) is presented in Fig. 12.

III.   CONTROL OF NITROGEN OXIDES

The formation of nitrogen (NO_x) oxides occurs from the
thermal oxidation of nitrogen present in the combustion air
and also from the thermal oxidation of nitrogen in the fuel.
The oxidation of nitrogen in the combustion air is known as
thermal NO_x while oxidation of nitrogen present in the fuel
or waste is known as fuel-bound NO_x formation. The formation
of nitrogen oxides, their emission rates for various types of
municipal and hazardous waste incinerators, and in situ control of NO_x were previously discussed. Thus, this section
focuses on post combustion control of NO_x, including noncatalytic ammonia injection, catalytic ammonia injection, and
flue gas recirculation.

A.   Noncatalytic Ammonia Injection

Noncatalytic ammonia injection is a process where ammonia
with a steam or air carrier is injected into a hot gas stream
containing nitric oxide. The ammonia reduces the nitric
oxide to diatomic nitrogen and water vapor. Noncatalytic
ammonia injection is applicable at NO_x concentrations ranging
from 40 to 10,000 ppm, with 300 ppm being typical [13]. Control efficiencies usually range from 40 to 70%. Typical
reductions on MSW incinerators in Japan range from 50 to 65%
at influent NO_x concentrations ranging from 80 to 160 ppm.
The following is a discussion of the system and its applications and efficiencies.

1.   *System Description*

Noncatalytic ammonia injection is based on the homogenous
gas phase reactions between nitric oxide (NO) present in the

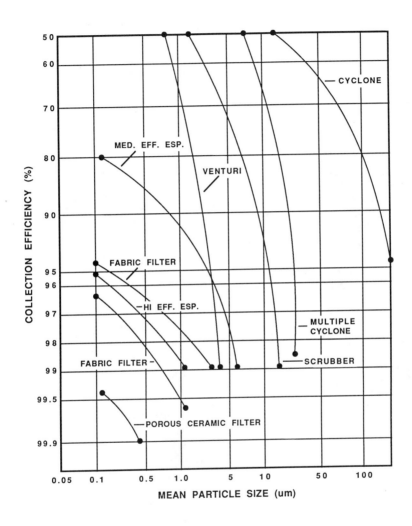

FIGURE 12.   Particulate Collector Performance.
Source: Zievers, Eggerstedt, and Kulousek [37].

flue gases from the combustion process and the injected
ammonia (NH_3). The products of this reaction are diatomic
nitrogen (N_2) and water vapor (H_2O). The ammonia is injected
into the gas path with a steam or air carrier which adds
momentum to the ammonia aiding its distribution into the

combustor. The following general reactions occur with
ammonia injection.

$$2NO + 4NH_3 + 2O_2 + 2H_2O => 3N_2 + 8H_2O \quad 1600-2000°F \quad (8-1)$$
$$(870-1095°C)$$

$$2NH_3 + 5/4O_2 + 1/2H_2O => NO + 2H_2O \quad >2200°F \quad (8-2)$$
$$(1205°C)$$

As can be seen by the above reactions, the temperature where
ammonia injection occurs is very critical. If the ammonia is
injected in the flue gas at a location above 2200°F (1205°C),
additional NO is formed. The optimum temperature range is
between 1600 and 2000°F (870 to 1095°C). Below 1600°F (870°C),
the ammonia passes through the combustor unreacted. The
optimum temperature range can be lowered to between 1200 and
1450°F (650 to 790°C) by injection of hydrogen (H_2) with the
steam or air carrier. The concentration of NO and NH_3 is a
function of temperature (Fig. 13) with the temperature window
being very narrow for ammonia injection with hydrogen. The
optimum temperature range is from 1250 to 1350°F (675 to
730°C) with hydrogen injection and between 1650 to 1850°F
(900 to 1010°C) without hydrogen injection.

The concentration of ammonia in the flue gas must be minimized before the gas exits the stack. Figure 13 indicates a
sharp rise in ammonia concentration as optimum operating conditions diminish. Excess NH_3 above that required for the
stoichiometric reaction with NO appears at the stack. This
excess NH_3 is called ammonia slip, or breakthrough, and is a
function of the desired collection efficiency and the influent
NO concentration in the flue gas [25]. More than 80% removal
efficiency is achievable at a molar ratio of NH_3 to NO of 1.0
when concentrations of NO in the flue gas are 200 to
400 ppm. However, when the concentration of NO in the flue
gas drops below 200 ppm, the molar ratio of NH_3/NO increase
to 1.5 to 3.0 for the same 80% removal. This is noteworthy
since the area of NO_x emissions of concern for newer MSW
incinerators ranges from 100 to 200 ppm of uncontrolled NO in
the flue gas. As a result, the amount of unreacted NH_3, or
ammonia slip, can increase significantly at these lower concentrations. Typical molar ratios applicable to MSW incinerators are betwen 0.5 and 2.0 for a collection efficiency of
40 to 70%. Typical NH_3 slip is approximately 20 to 50 ppm.
Average required gas velocities are about 4,000 feet (1,220 m)
per minute.

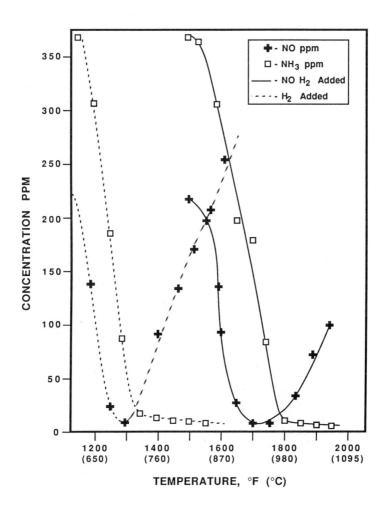

FIGURE 13. *Concentration of nitric oxide as a function*
of operating temperature for noncatalytic ammonia injec-
tion with and without hydrogen injection. Adapted from
Rosenburg et al. 1978. Control of NO_x emissions by stack
gas treatment, EPRI.

2. Applications

Noncatalytic ammonia injection has been used in Japan on
MSW facilities but has only recently been used in the United
States. Only a few installations exist in the United States
although many more are in the planning stage. Ammonia injection has gained rapid acceptability, particularly in

California where many areas are nonattainment areas for NO_x.
This is also important since NO_x is a precursor to the formation of ozone, and many parts of California are nonattainment
for ozone. In order to meet the regulations of the local air
pollution control districts, a process must remove around 40
to 70% of the NO_x. As indicated above, this has been shown to
be the case for many MSW facilities, mainly because the types
of large facilities burning MSW, 2,000 to 3,000 tons/day
(1,820 to 2,730 tonne) cannot effectively operate in a low
excess air or stage combustion mode.

For MSW applications the NH_3 and steam or air carrier
stream are injected into the furnace by the use of wall injectors. For proper selection of the optimum location of the wall
injectors, the furnace manufacture and the noncatalytic
ammonia injection vendor should be involved early in the
design. The optimum temperature area is usually before or
immediately after the superheat section (depending on boiler
configuration). As a result, the furnace must be configured
to achieve the appropriate residence times for the reaction
to occur. Also, multiple injectors are usually installed to
allow the continual operation of the NH_3 injection system at
reduced load, while still maintaining its designed efficiency.
The use of multiple injector locations eliminates the need for
hydrogen injection at reduced load conditions. The wall
injectors can be used in boiler retrofit applications as well.

There are no mobile, transportable, or fixed hazardous
waste facilities that use noncatalytic ammonia injection.
However, in hazardous waste incinerators it may not be as easy
to achieve the appropriate temperature range. For example,
the combustion gases exiting an afterburner (for example following a rotary kiln) may be in the temperature range of 2200
to 2300°F (1205 to 1260°C). As such, cooling of the gases
would be required. This could involve evaporative cooling or
cooling with air to 1600 to 2000°F (870 to 1095°C) or cooling
to 1200 to 1450°F (650 to 790°C) prior to the gas entering a
heat recovery steam generator (HRSG). Ammonia injection could
then be accomplished in the HRSG.

3. Corrosion Concerns

There is concern over corrosion of carbon steel surfaces
during startup and shutdown of MSW facilities. With lower
temperatures around 800°F (425°C) the sulfite ion (SO_3^-) in
the flue gas can combine with water to form sulfuric acid
(H_2SO_4). At lower temperatures around 600°F (315°C) ammonia,
in excess of the stoichiometric requirement for NO reduction,
reacts with the H_2SO_4 to form ammonium bisulfite (NH_4HSO_4) as
shown below.

$$NH_3 + H_2SO_4 => NH_4HSO_4 \tag{8-3}$$

As the temperature decreases further the NH_4HSO_4 further reacts with ammonia to form ammonium sulfate $[(NH_4)_2SO_4]$ as indicated below

$$NH_3 + NH_4HSO_4 => (NH_4)_2SO_4 \tag{8-4}$$

Another concern is the formation of ammonia chloride (NH_4Cl) at around 250°F (120°C) as indicated below.

$$NH_3 + HCl => NH_4Cl \tag{8-5}$$

Ammonium chloride is formed beginning at 190°F (88°C) and ending at 225 to 250°F (107 to 120°C) (Fig. 14). Other equilibrium reactions for a typical MSW fired incinerator are shown in Fig. 14. MSW is usually low in sulfur content and, therefore, the concentration of ammonia bisulfite formed is only around 2 ppm at 360 to 390°F (180 to 200°C) [14]. For a hazardous waste incinerator the levels of the sulfite ion and available chloride could be significantly higher. If a dry scrubber is used (cooling of the gas stream to 190°F or 88°C) the ammonia chloride will be formed and captured prior to a downstream baghouse or other equipment.

4. Efficiencies

Probably the best known commercially available noncatalytic ammonia injection system is the Thermal DeNOx system marketed by Exxon Research and Engineering Company. To date, 60 systems have been installed on industrial boilers, utility boilers, petroleum heaters, and oil field steamers. The process has also been tested in Kobe, Japan, on an MSW incinerator. In the United States, the Thermal DeNOx system has been successfully installed on a wood-fired power plant in Longbeach, California [13], and also at the 300 TPD (270 tonne) City of Commerce mass burn facility in Los Angeles County. Other systems are planned for MSW, sewage sludge, and wood-fired facilities [14]. Existing and planned Thermal DeNOx systems show typical removal efficiencies are 57 to 80% with ammonia slip at between 5 to 56 ppm (Table V).

B. Catalytic Ammonia Injection

In the selective catalytic reduction (SCR) process, ammonia gas is injected in either an air or steam carrier stream, through a grid, into the flue gas. The optimum temperature range for this system is between 600 to 800°F (315 to

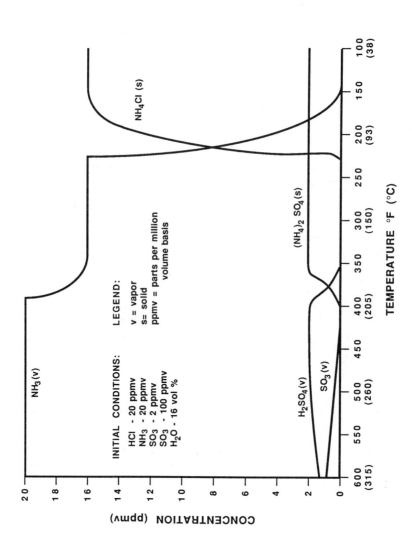

FIGURE 14. *Equilibrium reactions in cooled flue gases. Source: Hurst and White* [14].

TABLE V. Design and Operating Parameters for Thermal
DeNOx on Power Plants in California

Parameter	Facility 1	Facility 2	Facility 3	Facility 4
Fuel	MSW[a]	Wood	Wood	Sludge
Status	Operating	Operating	Operating	Under construction
NO_x initial ppmv	204	129	325	225
NO_x final ppmv	60	25	140	79
Reduction (%)	70	80	57	65
Ammonia slip	50	0	5-10	55
Injection temperature, °F (°C)	1750-1900 (955-1035)	1590-1960 (865-1070)	1600-1850 (870-1010)	1474-1550 (800-845)

[a]City of Commerce, California.
Source: Letter from B. E. Hurst [15].

425°C). The reaction mechanisms for NO and NO_2 are shown
below [17].

$$4NO + 4NH_3 + O_2 => 4N_2 + 6H_2O \qquad (8-6)$$

$$2NO_2 + 4NH_3 + O_2 => 3N_2 + 6H_2O \qquad (8-7)$$

The following is a discussion of the system and its applications and efficiencies.

1. *System Description*

In SCR systems gaseous ammonia is injected into the gas
stream at the appropriate location based on the system configuration. For example, it may be injected after the economizer in a large MSW incinerator. The ammonia and flue gas
mixture then travels to a catalytic reactor where the ammonia
and NO_x interact to form N_2 and H_2O [17]. The use of the
catalyst causes a sharp decrease in the temperature required
for the reaction to occur.

Important design considerations include the system configuration, selection of the molar ratio of NH_3 to NO, operating
temperature region, the space velocity, and the selection of
the catalyst as discussed below.

a. *System Configuration.* The system configuration
depends on the facility and any other required air pollution
control equipment. Configurations can include any of the
following: (1) boiler, followed by an ESP, the SCR, an
induced draft fan, a wet scrubber, and the stack; (2) boiler,
followed by an evaporative quench, a quench reactor for
removal of acid gases, a baghouse, the SCR, the induced draft
fan, and the stack; and (3) boiler, followed by a dry scrubber
(spray dryer), baghouse, flue gas reheat, the SCR, induced
draft fan, and the stack.
Key to each configuration is the need to keep the flue gas
as clean as possible to avoid blinding or poisoning the catalyst (caused by the presence of sulfur dioxide in the flue
gas). All of the above configurations would change the organization of the conventional MSW air pollution control equipment
train. The current trend favors a spray dryer and a baghouse
configuration, which would not be appropriate for the incorporation of SCR but would work for Thermal DeNOx (Noncatalytic ammonia injection).

b. *Molar Ratio.* For SCR systems the required molar ratio
of NH_3 to NO is approximately one. In actual practice the
molar ratio may be increased a little above one to achieve
an 80+ percent removal efficiency (Fig. 15). Lower molar
ratios can be utilized if only 40 to 70% or so removal is
required. In addition, the catalyst volume must be appropriately sized. If there is insufficient catalyst, the molar
ratio will need to be increased above 1.0, which would cause
an increase in ammonia slip.

c. *Temperature.* The ideal temperature range is between
600 and 800°F (325 to 425°C). Below this range ammonium sulfate and bisulfite may form and attack the catalyst. In a
typical MSW incineration facility an economizer bypass duct
may be installed to raise the economizer exit gas temperature
at low loads to an acceptable level and minimize corrosion.
If the temperature range is exceeded, the catalyst may be
sintered and permanently damaged.

d. *Space Velocity.* The level of NO_x removal is mainly
a function of the space velocity selected for the catalyst.
The space velocity is the total flue gas volume per hour
divided by the bulk volume of the catalyst. The units of

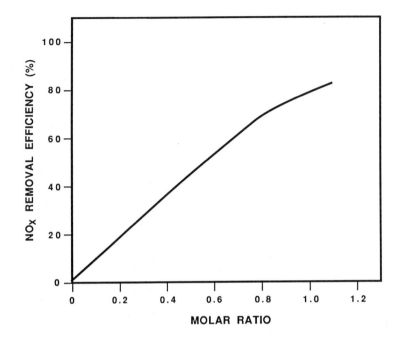

FIGURE 15. Relationship between nitrogen oxide removal efficiency and the molar ratio at the Tokyo-Hikarigaoka Municipal Incinerator using selective catalytic reduction. Source: Mitsubishi Heavy Industries, LTD [22].

space velocity are then reciprocal hours. The space velocity generally ranges from a low of 2,000 to a high of 10,000. Typical values used on MSW incineration systems in Japan have ranged from 3,000 to 5,500. The low end of this range represents increasingly large and more costly facilities, since a given volume of catalyst bed is handling less flue gas flow.

    e. *The Catalyst.* The honeycomb catalyst structure is used by Kawasaki and Mitsubishi while a plate type catalyst is used by Hitachi. It is claimed that the plate type catalyst is less susceptible to plugging. These structures are mounted parallel to the gas stream in the reactor in order to reduce the pressure drop and to reduce plugging. The most common catalyst is that of titanium dioxide-vandium pentoxide (TiO_2-V_2O_5). Upstream of the SCR unit, the ammonia is

vaporized with hot water or steam and injected into the duct.
The injection distance from the SCR is determined to maximize
good mixing before reaching the SCR.

2. Applications

As indicated above, the use of a catalyst lowers the temperatures at which the reactions occur when compared to noncatalytic ammonia injection. To date the major applications
of selective catalytic reduction (SCR) have been on boilers
firing heavy oil, liquefied natural gas, and coal, mainly in
Japan and Germany. Average efficiency of the systems has
ranged from about 30 to 90% NO_x reduction depending on the
plant [11, 16]. Systems installed in the United States (it
should be noted that of these systems installed, all but one
are in California) have included units on natural gas with an
average efficiency of 80 to 90% [11] and units installed on
coal and digester gas. Outside the U.S. more than 50 SCR systems have been installed in Japan, 13 in Germany, and 2 in
Austria on flue gas flows ranging from 6,000 acfm (170 m^3)
for a sludge incinerator to over 1,800,000 acfm (51,000 m^3)
for large central station plants.
    To date, SCR has not been used in the United States on
MSW or hazardous waste incineration facilities. However,
research and marketing are active in this area. Two SCR applications on MSW incinerators have been completed to date at the
Tokyo-Hikarigaoka and the Iwatsuki incinerators in Japan.

    a. The Tokyo-Hikarigaoka Incineration Facility. The
Tokyo-Hikarigaoka MSW incinerator has one unit with a capacity
per unit of 150 TPD (135 tonne) [22]. The process train for
this facility is a boiler followed by a hot ESP, the SCR,
induced draft fan, a wet scrubber, and the stack to atmosphere. The facility generates approximately 28,400 acfm
(805 m^3) of flue gas at 425 to 540°F (220 to 280°C). The
SCR utilizes 286 cubic feet (8.1 m^3) of catalyst and an
ammonia injection rate into the flue gas of 5.73 lbs/hr
(12.6 kg/hr). The inlet NO_x loading is 120 ppm with an outlet
loading of 60 ppm for a reduction of 50%. This unit became
operational on December 20, 1987.

    b. The Iwatsuki Incineration Facility. The Isatsuki MSW
incinerator has two units each with a capacity of 65 TPD
(60 tonne) [22]. The process train for this facility is a
boiler followed by an evaporative quench, a quench reactor
for acid gas control, a baghouse, the SCR, the induced draft
fan, and the stack to atmosphere. The facility generates
14,830 acfm (420 m^3) of flue gas at a temperature of 375 to

430°F (190 to 220°C). The SCR utilizes 286 cubic feet (8.1 m^3) of catalyst and an ammonia injection rate of 3.66 lbs/hr (8.05 kg/hr). The inlet NO_x concentration is 150 ppm with an outlet concentration of 30 ppm for a reduction of 80%. The units became operational on February 2, 1987.

The above facilities are small in size when compared to the 1,000 to 3,000 TPD (910 to 2,720 tonne) facilities planned and under construction in the United States. However, they are large enough to show that the SCR process is a viable alternative for the control of NO_x. The SCR process has not as yet found applications on hazardous waste incinerators.

c. Corrosion Concerns. The corrosion problems previously discussed are also a concern for the SCR process, particularly at startup and shutdown. Another concern is the deposition of ammonia bisulfite on the catalyst. The temperature of deposition is a function of the concentration of both SO_3 and NH_3 in the gas stream (Fig. 16)[11]. The low end concentration of NH_3 and the concentration of SO_3 controls the temperature of deposition, with the lower catalyst operating temperature better. However, operating temperature must be traded off with collection efficiency and catalyst life.

d. Efficiencies. As previously discussed, the collection efficiencies can range from 30 to more than 90% NO_x removal. Two applications on MSW incineration facilities have shown NO_x removal efficiencies of 50 and 80%. Higher levels of reduction can be achieved by simply increasing the catalyst volume. Outlet loadings as low as 4 to 9 ppm, representing a 95% reduction, are achievable [11]. However, these levels of control may not be desirable or cost effective. Factors which may affect the collection efficiency are the concentrations of sodium and potassium in the flue gas stream. At concentrations over 1.5%, collection efficiency is limited to about 60% for sodium and to 30 to 40% for potassium [11]. As operating time increases, the performance of the SCR decreases, and the molar ratio of NH_3 to NO must be increased. However, increasing this ratio increases the amount of ammonia slip. As ammonia slip reaches the regulatory limit, the catalyst must be replaced. Catalyst life is estimated to be about 2 to 5 years depending on the specifics of each application.

C. Flue Gas Recirculation

Another method of reducing the amount of NO_x in flue gas is flue gas recirculation. In this process, a portion of the flue gases is returned to the primary combustion zone of the

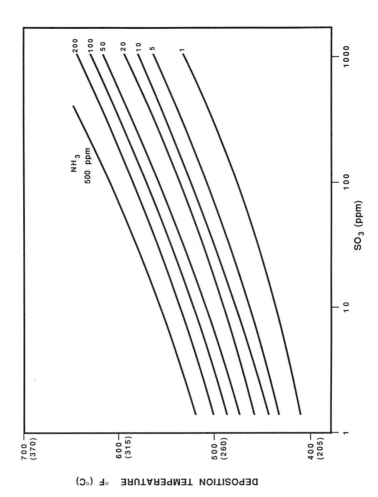

FIGURE 16. The concentration of SO_3 versus deposition temperatures.
Source: Hitachi [11].

furnace. These gases reduce the peak flame temperature and retard the formation of thermal NO_x. However, flue gas recirculation has not been effective at reducing the formation of fuel-bound NO_x.

Tests of flue gas recirculation have been conducted at the Kita refuse burning facility in Tokyo, Japan. During these tests, a NO_x reduction of 25% was achieved with a recirculation rate of 20 to 25% (Fig. 17) [10]. One concern with flue gas recirculation is that varying rates of recirculation could reduce the time-temperature relationship for proper destruction of products of combustion. As such, flue gas recirculation is of little use in controlling fuel bound NO_x emissions, unless only small removal rates are required.

IV. ACID GAS CONTROL

This section describes the technologies and efficiencies of systems used to control the following acid gases: hydrogen chloride (HCl), sulfur dioxide (SO_2), and hydrogen fluoride (HF). The technologies discussed include dry scrubbers (also known as spray dryers), dry injection (true dry scrubbing), and condensing type wet scrubbers. Also discussed are

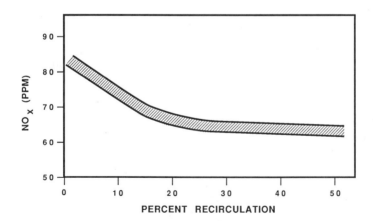

FIGURE 17. Effect of recirculation rate on nitrogen oxide emissions at the Kita refuse-burning facility in Tokyo, Japan. Source: Hirayama [10] as cited in CARB [4].

some of the current air pollution control equipment trains
and system configurations, since the selection of the scrubber
type establishes many of the other system components.

A. Dry Scrubbers

Dry scrubbers have become the most commonly specified acid
gas control technology for MSW incineration. Dry scrubbers
system operation, system descriptions, and system efficiencies are presented in the following section.

1. *System Operation*

In the spray dryer system the flue gases from the incinerator enter a spray dryer aborber. Into this gas stream a
finely atomized lime slurry (absorbent) is sprayed. The finely
divided lime spray reacts essentially quantitatively with the
acid gas in the flue gases. The principal reactions are summarized below for HCl and SO_2 [18].

$$Ca(OH)_2 + 2HCl \Rightarrow CaCl_2 + 2H_2O \qquad\qquad (8\text{-}8)$$

$$Ca(OH)_2 + SO_2 \Rightarrow CaSO_3 + H_2O \qquad\qquad (8\text{-}9)$$

Some carbonation of carbon dioxide (CO_2) occurs because of the
high concentration of CO_2 present in the flue gas. The reaction is shown below.

$$Ca(OH)_2 + CO_2 \Rightarrow CaCO_3 + H_2O \qquad\qquad (8\text{-}10)$$

The spray dryer has also been shown to be effective at the
removal of HF as well as the other acid gas components.

2. *System Description*

Major components of a spray dryer system include an absorber reactor, sorbent tank, reagent tank, fabric filter, I.D.
fan, and a flyash handling system (Fig. 18). The fabric
filter is used for almost all spray dryer operations since the
filter cake provides additional removal of acid gases. Probably the most common type of spray dryer is the rotary atomizer
which is manufactured by many domestic and foreign vendors.
A typical liquid-to-gas ratio may be on the order of 0.2 to
0.3 gallon/ft^3 (27 to 40 liters/m^3) of flue gas. The spray
dryer absorber is essentially a cylindrical vessel with a
cyclone shaped lower section. Flue gas enters at the top
where the lime slurry is sprayed into its path. The reactor
provides an approximate 10 to 12 second gas residence

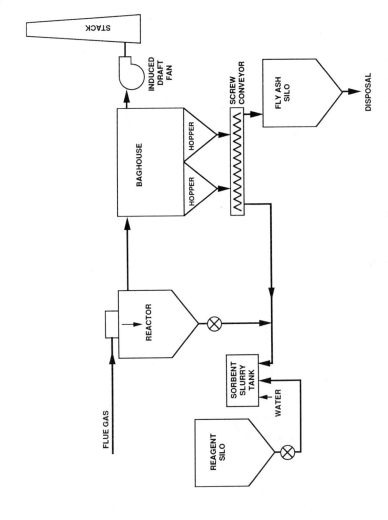

FIGURE 18. Typical spray dryer and particulate control system.

time [18]. The reacted solids (calcium chloride, calcium
sulfate, and calcium carbonate) exit at the bottom of the
reactor. The flue gases exit tangentially from the cyclonic
shaped lower section of the reactor. In some systems the
cyclonic separator can be a separate piece of downstream
equipment. Following this, the flue gas enters the fabric
filter and is ultimately exhausted to atmosphere. As indicated, most systems use fabric filters since an additional
10% removal can be obtained on the bag's surface. The most
common choice of bags are teflon-coated fiberglass.

The system can operate on flue gas directly from an
afterburner of a hazardous waste incinerator at between 2000
and 2300°F (1095 to 1260°C). However, operating history has
shown that it is desirable to cool the flue gases below 800°F
(425°C) to remove molten metals which can cause operating
problems. This quench can be done with water or air. The
flue gas temperature on MSW incineration facilities is typically below 500°F (260°C) before it enters the spray dryer
since the flue gas has passed over an economizer. With an
operating range of 400 to 800°F (205 to 425°C), the amount of
lime that can be accommodated is limited to an inlet concentration of 15,000 ppm HCl [18].

The solids formed from the spray dryer generally contain
less than 1% free moisture. Important design variables include the range of inlet concentrations, the desired outlet
concentration, the atomizer design, the necessity of flue gas
bypass for reheat, the sorbent preparation system, and the
design of the fabric filter.

3. *Efficiency*

Once-through SO_2 removal is a function of the reagent
ratio. The test conditions include a 300°F (150°C) flue gas
temperature, a 10-second gas residence time, and a 35°F (2°C)
approach; and results indicate that soda ash followed by
calcium lime are the most effective reagents. Proprietary
sorbents are also available for use in spray dryers. Sorbent
selection is very important, as are the approach to saturation of the flue gases in the reactor, and the degree of
reagent utilization. Typical curves of SO_2 removal and
reagent utilization as a function of the approach to saturation are presented in Fig. 19 for a system with a 1,600 ppm
SO_2 inlet concentration and a 2:1 Ca/S ratio [19].

The concentration of acid gases in the inlet to the spray
dryer affects the approach to saturation utilized. In typical
MSW incinerators the SO_2 concentration would be much lower,
maybe in the area of 200 to 400 ppm. Curves showing the SO_2
and removal efficiency as a function of the approach to

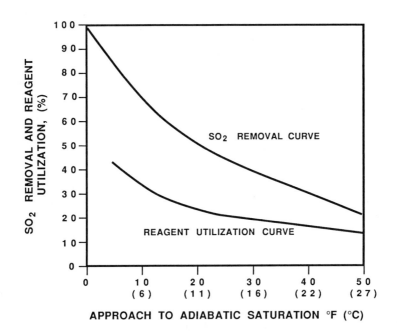

FIGURE 19. Effect of approach temperature on reagent utilization and sulfur dioxide removal for a 1600 ppm inlet concentration and a 2:1 calcium/sulfur molar ratio. Adapted from Luongo [19].

saturation is shown in Fig. 20 [3]. These curves were developed for a system with an inlet concentration of 276 ppm for HCl, 268 for SO_2, and 20 for HF. In all tests the removal efficiency of HF remained around 98.2%.

As mentioned above, research has shown that subsequent removal occurs in the baghouse on the bag surface. Selected results are indicated in Table VI for soda ash [24]. It can be seen at high removal efficiencies that sorbent efficiency decreases significantly. However, this can be offset somewhat by selection of the approach to saturation. Also on systems with low sorbent utilization, the sorbent can be recycled back from the baghouse.

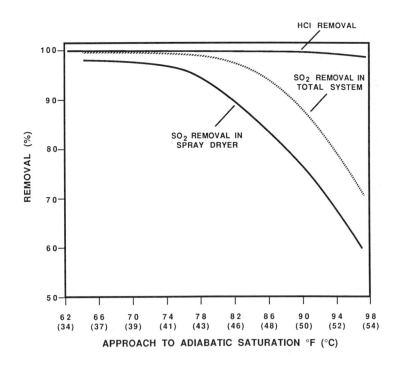

FIGURE 20. Sulfur dioxide and hydrogen chloride removal
efficiency versus approach to saturation using a 1.31 to
1.68 stoichiometric ratio and inlet concentrations of
250-280 ppm for sulfur dioxide and 90-160 ppm for hydrogen chloride and 3-35 ppm for hydrogen fluoride (note
that HF will follow the HCL curve). Adapted from Borio
and Plumley [3].

B. Dry Injection

In a dry scrubbing system the reagent is injected directly
into the duct without making a slurry. As a result the system
is totally dry. The system operation, description, and
efficiency are discussed below.

1. System Operation

Dry injection involves the pneumatic injection of a dry,
powdery sodium compound directly into the flue gas stream

TABLE VI. Percent SO_2 Removal as a Function of
      Stoichiometric Ratio

Stoichio-metric ratio	Spray dryer	Fabric filter	Total	Percent sorbent utilization
0.5	40	8	48	96
1.0	82	10	92	92
1.5	86	12	98	65

with subsequent downstream particulate removal in a baghouse.
Although many compounds have been tested, only certain sodium
compounds contained in such ores as nahcolite and trona appear
promising as well as cost effective. These ores contain
sodium bicarbonate ($NaHCO_3$), sodium carbonate (Na_2CO_3), and
sodium sesquicarbonate (Na_2CO_3, $NaHCO_3$, $2H_2O$). The major chemical reactions occurring for these compounds are presented
below [28].

$$2NaHCO_3 + SO_2 + 1/2O_2 => NaSO_4 + 2CO_2 + H_2O \qquad (8-11)$$

$$2(Na_2CO_3 \cdot NaHCO_3 \cdot 2H_2O) + 3SO_2 + 3/2O_2$$

$$=> 3Na2SO_4 + 4CO_2 + 5H_2O \qquad (8-12)$$

It has been shown that injection temperatures on the order of
275°F (135°C) are required for high collection efficiencies.

2. System Description

The process of dry injection includes a reagent pulverizer
and injection fan, storage of reagent in a bin, and use of a
fabric filter (Fig. 21). Prior to injection, the flue gas
may have to be quenched to around 270 to 280°F (132 to 138°C).
Cooling to the lower temperatures allows volatile vapors such
as metal chlorides, arsenic, mercury, dioxins, and other
organics to condense for subsequent absorption on the sorbent
and collection in the fabric filter [20]. This is also true
for the dry scrubber technology. The duct reaction zone
should be about 50 to 100 feet (15.2 to 30.5 m) long for a
uniform gas flow. Turning vanes and baffles are used as
required.

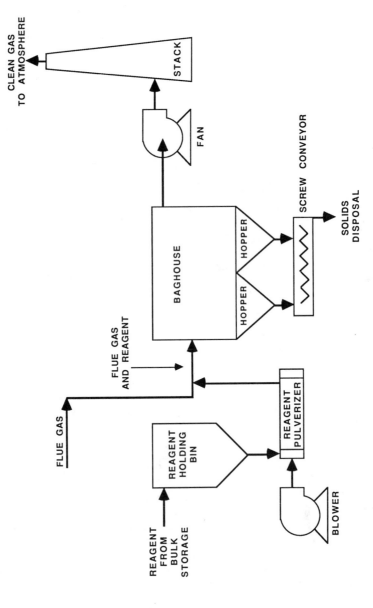

FIGURE 21. Flow diagram of a dry alkali injection system.

3. Efficiency

Dry injection is capable of meeting the collection efficiencies of dry scrubbing systems [20]. Also of interest in
dry injection systems is a 23% reduction of NO_x [20]. The
percent reduction of SO_2 as a function of the normalized
stoichiometric ratio for sodium bicarbonate and sodium sesquicarbonate is shown in Fig. 22 [12]. As indicated, efficiencies in excess of 90% are only possible at stoichiometric
ratios between 1.0 and 2.0. However, as the stoichiometric
ratio increases the sorbent utilization decreases. Sorbent
utilization at a stoichiometric ratio of 0.5 and 1.5 are
85% and 59%, respectively. Again, the importance of the
selection of the sorbent is obvious. Hydrated lime has been
used on a European kiln for the removal of SO_2, HCl, and HF.
Similar systems in West Germany have used calcium dehydrate
with steam injection. Efficiencies have been reported at
50 to 85% for SO_2 and over 90% for HCl for a 270 TPD (245
tonne) stoker-fired MSW unit [20].

The capital cost of a dry injection system is only 40 to
60% of a spray dryer system (both including the fabric
filter); however, operating costs of the dry injection system
are 40 to 60% higher than the spray dryer system mainly because of the low utilization of the higher priced pre-slaked
lime [9].

C. Wet Scrubbers

Wet scrubbers, used extensively on large central station
power generation facilities, have advanced both in operability
and in materials of construction over the years. Two major
problems associated with wet scrubbers are corrosion and
scaling at the wet-dry interface. These problems can be over-
come by using special construction materials and by limiting
the recycling of the scrubber solution. The wet scrubber
technology also requires a water treatment facility.

Of the total 190 MSW incineration plants in the construction, operation, or advanced planning stages, approximately
67 use or will use scrubbers [2]. Of these, 88% use dry
scrubbers (spray dryers) and 12% use wet scrubbers. Of the
8 wet scrubbers identified, 4 are at operating facilities,
2 at facilities under construction, and 2 at facilities in
the advanced planning stage. Thus, effluent-free systems are
clearly dominating the industry.

Typical wet scrubber systems require an extensive water
treatment system. However, for hazardous waste incinerators
spray evaporators are being used to evaporate the liquid and

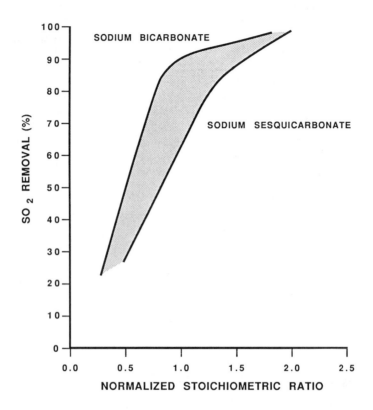

FIGURE 22. Sulfur dioxide removal as a function of
stoichiometric ratio for an in-duct injection system.
Adapted from Palazzolo and Bavieolo [28].

collect the solid residue in the wet scrubber effluent. This
type of system has been proposed on two fixed hazardous waste
incinerators. A typical system configuration following the
afterburner may include an air quench, heat recovery steam
generator (HRSG), spray dryer, baghouse or ESP, a water quench,
and a wet scrubber for removal of SO_2. Another alternative
being proposed is quench in a radiation chamber, HRSG, spray
dryer, baghouse, quench wet scrubber, and stack gas reheat.
Many other system configurations are available for hazardous
waste incineration facilities.

Wet scrubbers are finding use on hazardous waste facilities in the United States and Europe, where wet scrubbers cool
the flue gas to or below saturation. These units have the
advantage of high removal rates of metals, salts, and dioxins.
Wet scrubber based systems are discussed in the following section with system configurations.

D. System Configuration

The spray dryer and dry injection systems described above
are relatively straight forward in unit operations. However,
other systems have complex system unit operations, such as the
Teller, the Gotaverken, and the Flakt systems. Numerous systems are available which deserve equal evaluation, including
waste heat boilers for gas quenching, ESPs prior to acid gas
scrubbing, evaporative gas quenching, quench venturi, elec-
trostatically augmented wet scrubbers, and spray dryers for
zero discharge on wet scrubber systems [5].

1. *The Teller System*

The Teller Environmental Systems have been installed on
over twelve MSW and RDF incinerators ranging in size from
75 TPD (68 tonne) to over 350 TPD (320 tonne) [33, 34, 35].
The most recent installation was on two 275 TPD (245 tonne)
MSW units at the Marion County Facility in Oregon [26]. The
Teller system is basically a spray dryer system that uses a
dry venturi between the spray dryer and the baghouse. A
proprietary dry crystalline additive, called Tesisorb, is
injected into the venturi. The purpose of this additive is
to capture the fine particulates by inertial impaction, by
increasing the average particle size to over 10 microns [1].
A schematic representation of the Teller system is provided
in Figure 23.
    Total average particulate removal is on the order of 99.6%
[33]. Removal of particulate less than 17.5 microns and 1.6
microns ranges from 91.8 to over 98.7% [34]. Removal of
arsenic, beryllium, cadmium, chromium, nickel, zinc, lead,
copper, manganese, selenium, and tin ranged from 99.3 to 99.8%
[33]. Acid gases are removed to in excess of 99%.

2. *Gotaverken Condensing Wet Scrubber System*

The Gotaverken system, developed by Andeze AB of Heling-
burg, Sweden, is based on andeze's research on ultrafine
spray technology. A simplified schematic representation of
the system is shown in Fig. 24 [29]. Flue gas, which may
require evaporative cooling, enters a cyclone collector where

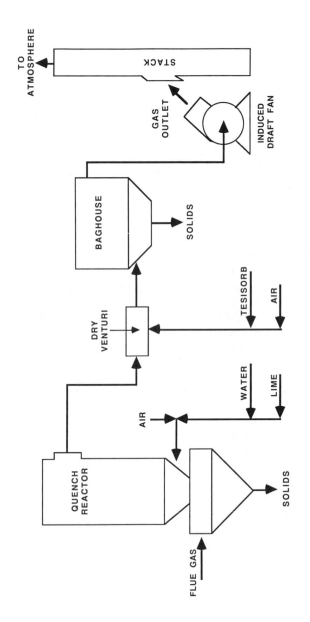

FIGURE 23. A simplified schematic representation of the Teller Environmental System technology. Source: [1].

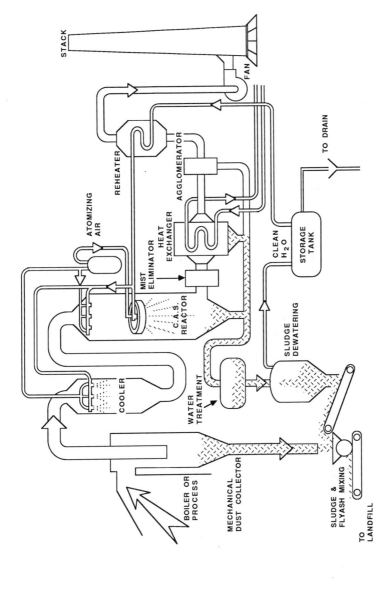

FIGURE 24. A schematic of the Gotaverken wet flue gas scrubbing system. Source: Raring Corp. [29].

331

over 80% of the particulate is removed (mainly greater than
10 microns). In the reactor vessel droplets of water react
with the acid gas components and dust particulate. Because
the flue gas velocity is lower, these droplets settle out in
the chamber and are drained. The gas leaves the reactor below
the saturation temperature and goes to a gas-to-water heat
exchanger where the temperature is reduced to 90°F (32°C),
recovering the latent heat of vaporization of the water vapor.
Water vapor and dust are then removed by a mist eliminator.
The flue gas is then reheated to 170°F (77°C) and exits
through the stack. The collected wastewater is neutralized,
heavy metals precipitated, and the pH adjusted. Sludge is
treated in a filter press and the heavy metals are stabilized
if required.

Particulate removal is similar to that of a baghouse
system: HCl removal exceeds 99%, SO_2 removal exceeds 98%,
HF removal exceeds 80%, mercury removal exceeds 92%, and
dioxin removal is in the range of 93 to 96%.

3. Flakt Process

Flakt manufactures a wide range of air pollution control
equipment, including its dry scrubbing system which has been
used on one facility in the United States. That facility,
however, only required a 70% removal of acid gases. Dry
injection systems have been installed at the Sysav incinerator
in Malmo, Sweden, and at the Hagdaken plant in South Stockholm,
Sweden. However, in the United States it appears that spray
dryers will be used where high acid gas removal efficiencies
are required. Flakt has spray dryer installations planned
at four MSW incineration facilities in the United States with
collection efficiencies of 90 to 96% [8].

V. CONCLUSION

The combustion of municipal and hazardous waste generates
substantial quantities of airborne emissions which, if uncontrolled, present unacceptable environmental impacts. These
airborne emissions include particulates which may be enriched
with heavy metals. They also may include acid gases such as
SO_2, HCl, and HF. The airborne emissions include organics
such as CO, hydrocarbons, and chlorinated dioxins and furans.
Finally, airborne emissions may include heavy metals in the
vapor phase, oxides of nitrogen, and other pollutants. All
such emissions are products of combustion.

It has been noted in previous chapters that alternative
combustion technologies produce varying quantities of airborne

emissions. Among the MSW incineration systems, mass burn
units tend to produce fewer particulates than RDF systems.
Further, they tend to produce reduced quantities of chlorinated dioxins and furans. Alternatively, however, mass burn
units have the potential to generate higher heavy metal
emissions than RDF combustion systems. Fluidized bed units
have the potential to minimize oxides of nitrogen, acid gases,
and dioxins and furans. Fluidized bed units incorporate emission control reactions within the combustion system itself.

Given the variety of wastes to be burned and the range of
combustion technologies available, air quality control systems have been developed to control the airborne emissions as
generated. Fabric filters or baghouses have become the
industry standard for particulate control, although electrostatic precipitators and hydrosonic scrubbers (particularly
for mobile and transportable hazardous waste kilns) are also
used for this purpose. Acid gases can be controlled by a
variety of techniques including wet scrubbers, dry scrubbers,
and dry injection systems. Dry scrubbers and dry injection
systems couple the injection of sorbent with baghouse capture
of the final product. Advanced systems include the Teller
design, which uses Tesisorb as the sorbent.

Oxides of nitrogen (NO_x) can be controlled by a wide
variety of techniques including ammonia injection, selective
catalytic reduction, and certain designs for ionizing wet
scrubbing. Alternative NO_x controls include flue gas recirculation and staged combustion, although these techniques cause
additional problems for waste incinerators. Organic emissions
such as dioxins and dibenzofurans can be controlled by condensing wet scrubbing systems and dry injection systems.
Other organic emissions are controlled by good combustion
practice. Heavy metals may be controlled by the temperature
of the post-combustion control system. Condensing wet scrubbers and ionizing wet scrubbers operated at low temperatures
are among the more effective techniques for controlling
heavy metals.

Many of the combustion technologies have been around for
several decades, if not generations. Rotary kiln technology
has a long and rich tradition. Even fluidized bed technology
has been around for some 50 years, having been employed for
coal gasification as early as the mid 1930s. What character-
izes the current generation of incinerators, then, is not only
the increased knowledge of combustion and combustion control,
but also the increased ability to control airborne emissions
once formed.

As knowledge of combustion has increased, the rate
of airborne emissions generated has decreased. As knowledge of airborne emissions has increased, the ability of

technology to control those emissions has dramatically improved. The combination of combustion knowledge and air quality control knowledge has led to incineration systems which meet the objectives of waste management and waste treatment in an environmentally acceptable manner.

REFERENCES

1. Beaumont Enviromental Incorporated. 1985. Meeting California Emission Requirements: Balancing Regulatory Requirements and Equipment Requirements. Prepared for the California Energy Commission, Sacramento, California.

2. Berenyi, E. 1986. Resource Recovery Yearbook, Directory and Guide. Government Advistory Associates, Incorporated, New York, New York

3. Borio, D. C. and Plumley, A. L. 1985. Dry Scrubber/ Baghouse Pilot Study on Control of Emissions from Incinerators. Presentation: The Joint Power Generation Conference, Milwaukee, Wisconsin, October 20-24.

4. CARB (California Air Resources Board). 1984. Air Pollution at Resource Recovery Facilities. Final Report. May 24.

5. Chao, C. C., and Eng, P. 1988. Assessment of Alternative Flue Gas Treatment Systems for Hazardous Waste Incinerations. Proc: The International Conference on Incineration of Hazardous, Radioactive, and Mixed Wastes. San Francisco, California, May 3-6.

6. Clark, M. J. 1987. Issues, Options and Choices for Control of Emissions from Resource Recovery Plants. Presented at the Sixth Annual Resource Recovery Conference, Washington, D.C., March 26-27.

7. EPRI. 1978. Economics of Fabric Filters vs. Precipitators. Electric Power Research Institute, Palo Alto, California, June.

8. Flakt. 1987. General Catalog and Lists of Installations. Flakt, Inc. Environmental Systems Division, Knoxville, Tennessee.

9. Foster, J. J., Hachhauser, M. L., Petti, V. J., Sandell, M. A., and Porter, T. J. 1987. Proc: Thermal Treatment of Municipal, Industrial, and Hospital Wastes. APCA International Specialty Conf., Pittsburgh, Pennsylvania, November.

10. Hirayama, N. 1975. Control of NO_x Emissions from Municipal Refuse Incinerators. Proceedings of the First International Conference and Technical Exhibit, Montreaux, Switzerland, November 3-5.

11. Hitachi America Ltd. 1987. Presentation on Selective Catalytic Reduction. California Energy Commission, Sacramento, California, July 1.

12. Hooper, R. 1987. Full-Scale Demonstration of Dry Sodium Injection at R. D. Nixon Unit 1. ECS Update, Environmental Control Systems Department, Coal Combustion Systems Division of the Electric Power Research Institute, Palo Alto, California.

13. Hurst, B. E. 1986. Personal Communication. Exxon Engineering Technology Development, Materials and Heat Transfer Division, Florham, New Jersey, July 10.

14. Hurst, B. E., and White, C. M. 1986. Thermal DeNOx: A Commercial Selective Noncatalytic NO_x Reduction Process for Waste-to-Energy Applications. Paper presented at the 1986 National Waste Processing Conference, Twelfth Biennial Conference Waste Reduction – Conservation. ASME Solid Waste Processing Division, Denver, Colorado, June 1-4.

15. Hurst, B. E. 1987. Personal Communication. Exxon Engineering Technology Development, Materials and Heat Transfer Division, Florham, New Jersey, April 30.

16. Kawasaki. 1987. Presentation by aerequipment engineers at Ebasco Services Incorporated, Seattle, Washington, August 8.

17. Kerry, H. A., and Weir, A. 1982. Catalytic DeNOx Demonstration System Huntington Beach Generation Station Unit 2. Paper presented at the EPRI 1982 Joint Symposium on Stationary Combustion NO_x Control, Vol. 1, Utility Boiler Applications.

18. Kroll, P. J., and Williamson, P. 1986. Applications of Dry Flue Gas Scrubbing to Hazardous Waste Incineration. *Journal of the Air Pollution Control Association 36*(11: 1258-1263.

19. Luongo, S., Emmel, B., Offen, G. R., and Stern, R. D. 1986. Proc: 1986 Joint Symposium on Dry SO_2 and Simultaneous SO_2/NO_x Control Technologies; Vol. 1, Sorbens, Process Research, and Dispersion; Vol. 2, Economics, Power Plant Integration, and Commercial Applications. Prepared by Radian Corp, Research Triangle Park, N.C. Prepared for U.S.E.P.A. and the Electric Power Research Institute.

20. Makansi, J. 1986. New Processes Enhance the In-Duct Emissions-Control Option. *Power 130*(7):27-29.

21. Means, J. D., and Holland, O. L. 1986. Utilization of Hydro-Sonic Scrubbers for Abatement of Emissions from Hazardous/Low Level Radioactive Mixed Wastes. Proc: The Conference of Incineration of Low-Level Radioactive and Mixed Wastes, Charlotte, North Carolina, April 22-25.

22. Mitsubishi. 1987. Mitsubishi SCR System for Municipal Refuse Incinerator Measuring Results at Tokyo-Hikarigaoka and Iwatsuki. Mitsubishi Heavy Industries, Yokohama Dockyard and Machine Works.

23. Moore, E. B. 1984. Control Technology for Radioactive Emissions to the Atmosphere at U.S. Department of Energy Facilities. Pacific Northwest Laboratory, Richland, Washington.

24. Morasky, T. M. et al. 1977. EPRI's Flue Gas Desulfurization Program, Results and Current Work. Presented at the EPA Symposium on Flue Gas Desulfurization, Hollywood, Florida, November 8-11.

25. Muzio, L. J., and Arand, J. K. 1976 Gas Phase Decomposition of Nitric Oxide in Combustion Products. Sixteenth International Symposium on Combustion.

26. Ogden Projects Inc. 1986. Marion County Solid Waste-to-Energy Facility Boilers 1 and 2 Air Emissions Source Test Report. Prepared for Oregon State Department of Environmental Quality, Salem, Oregon.

27. Ostop, R. L. 1982. Update on Baghouses in Utility Operations. Presented at Rocky Mountain Electrical League Spring Conference, Cheyenne, Wyoming, May 4.

28. Palazzolo, M. A., and Baviello, M. A. 1983. Status of Dry SO_2 Control Systems. Prepared by Radian Corporation, Research Triangle Park. Prepared for the Industrial Environmental Research Laboratory, Research Triangle Park.

29. Raring Corporation. 1986. Presentation on the Andeze Process at Ebasco Services, Seattle, Washington, August 6.

30. Rousseau, P. 1987. The Tubular Electrostatic Precipitator for Control of Incinerator Flue Gas Emissions. Proc: Thermal Treatment of Municipal, Industrial, and Hospital Wastes. APCA International Specialty Conference, Pittsburgh, Pennsylvania, November.

31. Schneider, G. G., Horzella, T. I., Cooper, J., and
    Striegl, P. J. 1980. Selecting and Specifying Electro-
    static Precipitators. *In* Industrial Air Pollution
    Engineering. (V. Cavaseno, ed.), McGraw-Hill Pub. Co.,
    New York, New York.

32. Shah, J. T. 1987. Application of Wet Tubular Electro-
    static Precipitator to Control Emissions from Waste
    Incinerators. Proc: Thermal Treatment of Municipal,
    Industrial, and Hospital Wastes. APCA International
    Specialty Conference. Pittsburgh, Pennsylvania,
    November.

33. Teller, A. J. 1985. Teller System Incineration –
    Resource Recovery Fluid Gas Emission Control. Paper
    presented at the Acid Gas and Dioxin Control for Waste-
    to-Energy Facilities, Washington, D.C.

34. Teller, A. J. 1984. The Landmark Framingham, Massachu-
    setts, Incinerator. Paper presented at the Hazardous
    Materials Management Conference.

35. Teller, A. J. 1978. New Systems for Municipal Incinera-
    tor Emission Control. Paper presented at the 1978
    National Waste Processing Conference.

36. Western Precipitator. 1978. Type "V" Turbulaire
    Variable Venturi Scrubber. Brochure S-2.

37. Zievers, J. F., Eggerstedt, P. M., Kulousek, E., and
    Zievers, E. C. 1987. Porous Ceramics Modified for Hot
    Gas Cleanup (HGCU). Proc: Twenty-Fourth General Meeting,
    American Society of Sugar Beet Technologists. Phoenix,
    Arizona, March 1-5.

INDEX

Grate stoichiometric ratio (GSR), 66, 72, 75,
    96, 140
Grates,
    materials of construction, 71
    types of, 64, 74, 89

H

Hartford, Connecticut, 113, 117, 122–124, 138
Hazardous waste characteristics, 11
    corrosivity, 11
    ignitability, 11
    reactivity, 11
    toxicity, 11
Hazardous waste chemicals, 5, 12, 160,
    162–175
    empirical formula, 164
    heating value, 168–169, 171
    molecular weight, 164
    reactivity, 171–175
    structure, 165
    ultimate analysis, 168–169, 171
Hazardous waste generation, 5, 11–12
Hazardous waste generators, 12
Hazardous waste laws. See also specific laws
    Comprehensive Environmental Response,
        Compensation and Liability Act
        (CERCLA), 14, 15–16, 157, 159, 202,
        245
    Resource Conservation and Recovery Act
        (RCRA), 14, 160, 201
    Superfund Amendments and
        Reauthorization Act (SARA), 15, 202
    Toxic Substances Control Act (TSCA), 14
Hazardous waste reduction, 1
Hazardous waste treatment technologies,
    13–14
    biological, 13
    chemical, 13
    neutralization, 1
    oxidation-reduction, 1, 13
    stabilization, 1, 13
Hazardous waste types, 11
    liquids, 11, 162, 209
    sludges, 11, 162, 209
    solids, 11, 162, 209
Heat and material balances, 69, 228, 229
Heat release rates
    grate heat release, 67, 75, 81, 99–101, 102,
        103, 141

volumetric heat release, 67, 75, 81, 87,
    99–101, 102, 103, 141, 150, 221, 238
Heavy metals, 33, 119, 127, 139. See also
    Airborne emissions
Hemicelluloses, 25
HIM, 157, 212–213
Hogdalen, Sweden, 63, 64–69
Household hazardous wastes, 32

I

Incineration technologies. See Combustion
    technologies
IT Corporation, 253–258

J

JANAF Thermodynamic Tables, 43
Japan, 3

K

Kommunekemi A/S, 157, 203, 212
Kovik, Sweden, 117, 128–134

L

L.C. Steinmuller, 88, 89
Landfill costs, 6
Landfill life extension, 21, 135
Landfilling, 1, 6, 10, 132, 135
Landfills
    hazardous waste, 3, 14
    municipal waste, 2–3
Liquefaction, 1

M

Madison, Wisconsin, 117, 128, 129
Martin GMBH, 64, 74, 77, 89, 91
Mass burn facilities, 21
McKay Bay Facility. See Tampa, Florida
Metals, 4, 7. See also Airborne emissions
MSW-to-energy systems
    components of, 59
    costs of. See Costs
    existing facilities, 60
    vendors of, 60, 64